Brain, Nerves, Muscles and Electricity

My Life in Science

Brain, Nerves, Muscles and Electricity

My Life in Science

WLADIMIR THEODORE LIBERSON, MD, PhD

Compiled and Edited by

ROBERT COHN, MD

CATHRYN W. LIBERSON, PhD

Hist.
QP341
L47
1999

SMYRNA PRESS

1999

© Copyright 1999
by
SMYRNA PRESS

All Rights Reserved

Library of Congress Catalog Card Number 99-070993

ISBN 0-918266-38-6

No part of this book may be used or reproduced in any manner without written permission except for purposes of evaluative reviews. No part of this book may be stored in a retrieval system or transmitted in any form or by any means, including electronic, magnetic tape, mechanical recording, or otherwise without the prior consent of the publisher.

For information:
Smyrna Press, Box 1151, Union City, NJ 07087
(201) 617-7247

For permission to photocopy:
Copyright Clearance Center, 222 Rosewood Drive
Denvers, MA 01923
(508) 750-8400 or Fax (508) 750-4743

Printed in the United States of America
by
ATHENS PRINTING COMPANY
337 West 36 Street
New York, NY 10018-6401

Table of Contents

PREFACE *by Cathryn W. Liberson, PhD* — 7

PROLOGUE *by Oliver Sacks, MD* — 9

INTRODUCTION *by Robert Cohn, MD* — 11

Chapter 1 RUSSIA (KIEV, MOSCOW, LENINGRAD) — 15

Chapter 2 FRANCE (SORBONNE, SALPETRIERE) — 35

Chapter 3 U.S.A. (NEW YORK, HARTFORD) — 55

Chapter 4 PSYCHIATRY — 73

Chapter 5 HIPPOCAMPUS — 87

Chapter 6 NERVE CONDUCTION, EVOKED POTENTIALS & EMG — 101

Chapter 7 HUMAN LOCOMOTION — 131

Chapter 8 MICROSURGERY — 153

Chapter 9 FUNCTIONAL ELECTRICAL STIMULATION — 169

Preface

FIRST OF ALL I would like to express my deep gratitude to Dr. Robert Cohn for his truly heroic work in compiling this book which I never could have accomplished on my own. I also want to thank my stepdaughters: Annette, for her transcribing, translating and other input; and Helene, for her suggestions and constructive comments.

During the last thirty years of his life, my husband worked sporadically on his memoirs; but his reluctance to give up daily practice and interaction with patients prevented him from organizing the material into a final manuscript. He told me that he hoped it would be completed after his death if he was unable to do so. With Dr. Cohn's devotion to this project his hope is being realized.

My personal regret is that some of his writings were lost in the year following his death in 1994. He was adamant about never throwing anything away and consequently had amassed a staggering amount of records and papers; enough to fill three large warehouse rooms plus large amounts of material in his offices and at home.

He was undecided about whether to target the book to the scientific community or to the general public and therefore wrote copious notes with both in mind. Dr. Cohn and I have included some if not all of both.

He was also ambivalent about finishing the book at all. When his daughters or I pressed him to complete the work, he would say: "Why should I? Who cares whether I or someone else was the first to discover something; the important thing is that the discovery was made." However, as he was nearing his eighty-fifth birthday he convinced himself that it might be worthwhile after all. At that time he wrote that it was the joy of discovery; the unveiling of the unknown that has intrinsic value and to communicate this to others who follow that road justifies the effort of writing a book.

Until the end of his life he remained fascinated and excited by the meaning of multiple distinctive brain rhythms. When he formulated his

"Law of 3.5" in 1938 he hypothesized that one of the main functions of these rhythms is to decipher the temporal analysis of all incoming information. He stated that he believed that it would only be a professional neurophysiologist who would some day explain the need for these various rhythms.

Cathryn W. Liberson

Prologue

ALTHOUGH I HAD KNOWN of Dr. Liberson's pioneering work when I was a medical student in the mid-fifties, I only met him towards the end of his life when I consulted him as a patient with some nerve and muscle problems. Despite his age and frailty I found him brilliant, articulate and utterly charming. He did an examination which I can only call exquisite, from its delicacy, its detail, and the eager intelligence with which it was performed. There was nothing standard, nothing mechanical about it; one felt his mind considering, questioning, focusing, testing, honing in with every question, every observation, every test. EMG's and nerve conduction tests, for him were an integral part of the clinical examination, and he would pass from questions, to neurological testing, to inserting electrodes, and back, one to another without a break, as he delineated the precise shape and nature of the problem. It was beautiful, intellectual enquiry and sculpture in one.

He was intuitive, an artist in his work, no less than a minute analyst. And as he examined me; the examination like all of his examinations lasted two hours or more, he found me an eager listener. He reminisced and spoke of his early days as a young man in Russia and his more than sixty years of research into the electrical activity of brain, nerve and muscle.

When he had defined my problems; partial denervations of many muscles, and the causes of these; he prescribed one of the tiny, powerful electric stimulators he had invented for stimulating contractions in denervated muscles, a device I found invaluable during the months when I still had significant denervation.

This was in late 1993. Soon afterwards, he was taken ill. His office, full of books, manuscripts, ongoing work, photographs and mementos of an extraordinary life and a life still vitally creative in his ninetieth year, stood empty for a week, a month—and then we realized he would never come back. I think I was one of the very last patients he saw.

Oliver Sacks, MD

Introduction

SEVERAL MONTHS after the death of Dr. Wladimir Theodore Liberson in 1994, I was advised by his wife, Cathryn and his daughter Annette, that he had unfinished manuscripts which were to distill and express his ideas, interests and published work resulting from nearly seventy years of intense study in the neurological sciences. Because of my long standing friendship and general admiration of his ideas and productivity I suggested that Cathy should send some of the material to me for study.

I first met Volia just after WWII, in Hartford, Connecticut at the first meeting of the Eastern EEG Society. We had communicated by phone numerous times when he and Chick Stevenson were organizing the Society. This organization was created to stimulate training and research in electroencephalography in the United States.

Volia was the first clinical electroencephalographer in France and came to this country to escape annihilation by the evil forces operating in his adopted country. He was a medium tall, blonde man who spoke in a tolerant friendly way with only a moderate foreign accent. I was still mobilized in the U.S. Navy and he showed me his newly acquired instrumentation which was duly admired. We immediately began to talk about our common interests and a friendship was established until his death. At the various national and international meetings he always sat in the first row avidly listening and cogently (and sharply) discussing the presented material. He was very incisive in his remarks, but elicited no anger.

At dinner he was a delightful raconteur, and a real gourmet; we looked forward to our numerous re-unions. It was always a pleasure to argue with him about his ideas of the role of the EEG in brain function, which were never ordinary. I last saw him alive in Miami and he said, "Bobby, I am very sick."

What I received from Cathy was a massive amount of transcribed notes and numerous handwritten notebooks filled with his scientific writings. Some of the data were relatively complete and precise, but the actual printed material was quite limited, and much had been transcribed literally from sound tapes. Thus, for the most part I used his notebook data. This was in a Russian cursive form that I initially found difficult to read, but with perseverance this problem was overcome and a real order became manifest.

Especially trying were his variations in spelling the Russian names of significant persons. An example is afforded by the name of the great physiologist, Sechenov (1820–1905); in various places it was spelled Setchinow, Setchinov, Setchinoff and Sechenow. For Beritoff, who was a prominent contemporary physiologist, he occasionally used the Georgian name that I have omitted.

We have partly altered Volia's outline for the chapters; the last chapter that he proposed was to be "poetry and epilogue." Instead, we have put at the head of each chapter, appropriate insightful quotations from his "Eugene Onegin Revisited" which he published in 1987. These quotations nicely portray his literal poetry. The epilogue component he alone could have supplied.

We believe that the final product of his writings is the sensitive, almost poetic, story of a man completely dedicated to the understanding of the biological behavior of man, as well as his delightful guinea pigs and his sturdy rats. Throughout the dissertation he was a real physician with treatment as a major goal; the relief of pain and disability of his patients was always paramount.

The material that his wife and I have compiled in a direct coherent way, without distorting or changing his ideas or contents, is presented. How well we have accomplished this project must be left to the projected reader.

Robert Cohn, MD

Chronology

1904–1924	Russia (Kiev, Moscow, Leningrad)
1924–1940	Paris, France
1940–1941	New York City, New York
1941–1959	Hartford, Connecticut
1959–1971	Chicago, Illinois
1971–1973	Miami, Florida
1974–1981	Brooklyn, New York
1981–1990	Norfolk, Virginia
1991–1993	New York City, New York
1993–1994	Miami, Florida

CHAPTER I

Russia

*"At first Madame took care of the boy,
Then Monsieur tried to be his savior;
But very early Gene's whim and joy
Were the sole guides of his behavior"*

MY FATHER WAS one of 14 living children. He was a practising physician in a village near Kiev when, with the help of his family, he received training in the new science of medical radiology in Munich, Germany. He later moved to Kiev where he established a prosperous radiology diagnostic clinic and I was born there in 1904. My mother was an obstetrical nurse. My grandmother seemed to have been the strong provider in my father's family. Her husband was very religious and was occupied by reading the Talmud, etc., while she had to make the living. She had 20 children, of whom 14 survived and 6 received higher education, one daughter and 5 sons. (Obviously with this progeny her husband was not always studying religion.) Her first business venture was to select peasant girls in the village with long tresses whom she "sheared" and sold their hair in Kiev; it was a common practise for orthodox women to wear wigs. When in Kiev she observed the need of the Kievians for lingerie. She established a sort of factory at home. The next step was to rent a little store where she hired girls from the town, and despite the fact that she had 12 children to care for, she succeeded to organize the factory. At this time she moved her family to Kiev. My father was the eldest son and practising medicine as noted above. My grandmother had diabetes mellitus but she was strong enough to go to Paris to see the latest clothes models, which she copied in her factory. My grandfather took time from praying to keep the books for the factory.

I didn't go to grammar school. We had frauleins, my sister and I. Anti-Semitism in Russia was very ugly but my family was more or less spared from some of the undesirable aspects. For example, Jews were not allowed to live in certain parts of the city or on certain sides of the street, but physicians were exempt from this interdiction. I was the only Jewish student in a class of about forty and suffered from daily corporal attacks. My father, however, had the conviction that one cannot retreat so I had to continue to assert myself.

In the high school which I attended, I was subject to many indignities because I was told that I killed Christ. Remember that the revolution did not start in October but in February of 1917 and between February and October the people were free. In January of 1918 there was a constitutional assembly in which Bolsheviks were in a small minority.

Also I often hear today of a curious misunderstanding of history. For example, we hear that the name of Petersburg was changed in 1914 because of the war with Germany. In fact, Petersburg became an unpopular name and was changed to Petrigrad and only after the death of Lenin became Leningrad. People also felt that Lenin was a more acceptable dictator. In fact, Lenin was the originator of the notion of the dictators' regime and when he was asked what to do with the Czar's family, he did not hesitate to agree that all the family had to be killed.

In the spring before my birthday and in the fall after my bar mitzvah, Russia was convalescing in the throes of two revolutions bringing social transformation. As a Jew I could not but hate the Czar. It was he who condoned, if not instigated, pogroms against the Jews. It was his government and his church who were responsible for my miseries at school and my physical sufferings as well. It was he who represented the clan condemned throughout the country by the best Russian writers: Pushkin, Tolstoy, Gorky and Block.

In 1917, at the age of 13, I by chance attended a meeting of Russian Mensheviks in my native city of Kiev. After the revolution in February, there were many thousands of overjoyed people marching in the streets and so I was immediately taken by the idea of democracy, but after the October revolution, things changed. I was walking in the streets one night and saw people congregated. I joined them and saw a girl at a desk and asked what the meeting was about. She said they were Mensheviks and suggested that I should join the party. I said, "Don't you think I'm a little young for that?" She answered, "No, we have a Menshevik Youth. Are you sixteen?" I said, "About," and that is how I joined the Menshevik Party. Thus at that fateful meeting I joined the Menshevik Youth and later became a target for communist persecution. What attracted me to the Mensheviks was the idea of democracy, protection of workers and supremacy of science. I was not active politically but attended the meetings because of the very high quality of the men who were in the Party.

Because my father was a physician and my mother was an obstetrical nurse it was therefore natural for me to become a physician, which was the most ardent wish of my parents. However, the cruelty of the Czarist regime, the injustice of Russian anti-Semitism, the promises of the great February revolution and the greatness and liberalism of Russian literature made me hesitate many times between devoting myself to science and medicine and the compelling duty toward social and humanistic pursuits for the improvement of the lot of the common man.

I was not an outstanding student but I excelled in subjects of interest to me, particularly Russian and world literature. I was an average student of mathematics; the symbols of algebra left me unmoved and geometry seemed unrelated to my needs. Many years later I bitterly regretted the lack of my mathematical education.

When the time came for me to enter the University I continued to debate whether I should become a physician, a social scientist or perhaps even an electrical engineer. I passed the entrance examinations to all three institutions of higher education; the Medical School of Kiev, the Commercial Institute (which was devoted to social and economic studies), and to the Kiev Polytechnical. I chose the Kiev Medical because I thought it offered more important notions about man and his destiny. But I was not very happy because my memory was not so good and I didn't like to memorize all the Latin terms for anatomy. But my adventures in the Kiev Medical School were short-lived as the whole family moved to Moscow when I came under scrutiny by the police.

At the Medical School of the University of Moscow I continued my medical studies without great enthusiasm for the same reason that I had in Kiev. More than ever, I questioned the wisdom of my decision to go into medicine instead of becoming involved with the discussions of philosophy of the rationality of the different systems of national economy. Despite my increased doubts I persevered in medical studies, until once again the police began a search for me. At that time I arranged a transfer to the University of Leningrad, but because of my disinterest in anatomy I enrolled in classes of physiology at The Faculty of Science instead of at The Medical School. Another reason for the choice of the Faculty of Science was that Pavlov was Professor at the Military Medical Academy and it would have been impossible for someone on the blacklist of the secret police to become a matriculated student in a military school. However, I was able to enroll unobtrusively, as an unmatriculated student, at The Military Medical School to audit the lectures of the celebrated Pavlov. I was 18 at the time.

The Faculty of Science was very famous because one of the most outstanding early Russian physiologists, Sechenov, had been a Professor at the school. He was a legendary figure because he was one of the first Russian biologists who decided to study the physiology of the nervous system. He was privileged to be able to go abroad and work in the laboratory of Claude Bernard where he made a discovery, simple as to its technical achievements, but pregnant in its incalculable consequences, clearly foreseen by its discoverer. Sechenov experimented with the so-called "Turk" reflexes of the frog. The technique of the elicitation of the Turk reflex was very simple; a decapitated frog was suspended over a glass filled with weak sulphuric acid and one of its legs was plunged into the solution; the number of seconds that elapsed for the time of reflex withdrawal was measured. Sechenov varied the experiment by retaining some brain structures. Then without great sophistication he would take a crystal of "salt" and put it in the upper aspect of the remaining brain stem. To his amazement he observed that in some preparations the latency time for the withdrawal was considerably prolonged. He correctly interpreted this phenomenon by identifying it with an active process of inhibition originating from the brain stem and acting on the reflex centers of the spinal cord. Prior to Sechenov, central inhibition was not entertained; inhibition had only been described for heart action resulting from vagus nerve stimulation. Not only did he describe this phenomenon, but he carefully outlined the areas of the brain stem from which he obtained inhibition by electrical stimulation, or from the application of salt. He also described stimulation of other areas of the brain stem which induced a facilitation process as manifested by a shortened latency of the withdrawal response. And so Sechenov, in the middle of the last century, described clearly the effects that were rediscovered, in mammals, in this century by Magoun and his co-workers in the functional operation of the ascending and descending reticular formation of the brain stem.

Despite the greatness of the scientific work of Sechenov the reason for his profound influence on the thinking Russian intelligentsia was his conception that all behavior of man was due to responses of the brain to environmental stimuli, or to their traces stored in memory. Of course this idea is not revolutionary to us at the present time, although some of our philosophers may still not subscribe to it, but if one imagines the religious society of Russia in the 1860's when Sechenov's "Reflexes of the Brain" first appeared, one can understand how daring, how profoundly revolutionary this idea was at that time. Moreover, this monograph was written in such a convincing literary style, with such infectious enthusiasm, with such a militant call to abandon traditional ways of thinking, that it left its imprint on the cultural society of Russia.

Because of these revolutionary ideas Sechenov could not be endured by the Czar and his associates. As a consequence he was removed from Chairs of various universities; however this only increased his influence on Russian society. For the first time a physiologist had a dominant position in the intellectual life of the thinking people.

Sechenov's book argued that mental activity could be explained by a succession of reflexes, the endpoint would be muscle contraction or secretions. He said that emotional life could not be understood unless one took into consideration that muscles are involved in these manifestations. He also promulgated the tremendously profound idea that the environment is a fundamental constituent of organismal life. For example, when one cries there is activity in the lacrimal glands and the tears are squeezed out; likewise when one laughs, it is the muscles of the mouth that are involved (together with the respiratory apparatus); but with these actions an environmental happening set-off the laugh, etc. He claimed that there is no single emotion that does not have an end-point in muscle activity. But above all, his materialistic approach led him to the concluding concept that all men are equal, and that the most backward nationalities could be raised to a high cultural level by means of instructive education. Such ideas were anathema to the Czar and his censors.

After he wrote this daring controversial book on brain function he was forced to move to Odessa. His vacated Chair was contested by Wedensky and by Pavlov. In the contest Wedensky won and Pavlov was made a Professor in the Military Medical School.

When I came to Leningrad Wedensky had been dead for a couple of years, but the laboratory was still penetrated by his teaching, his spirit and his equipment. Wedensky was well known and quoted in all classical textbooks of physiology as responsible for "Wedensky's Inhibition," the observation of inhibitory action resulting from stimuli too strong or too frequent. To those in Physical Medicine and Rehabilitation he should be remembered as "the first electromyographer." As soon as the telephone was invented, he modified it so that he could listen to "currents" originating in the muscle. He demonstrated this "electromyophone" at several international physiological meetings. When his Chair was open for competition two of Wedensky's students, Prince Uchtomsky and Beritoff, a man from Georgia, were nominated. Uchtomsky won and Beritoff went to Tblisi, where he became a Professor of Physiology.

Uchtomsky developed his own theories of the activity of the central nervous system. At that time (1922) the University was not heated and Professor Uchtomsky, who had a very long beard like Tolstoy, wore his beard inside his overcoat when he lectured to us, a thing I can never forget. Anyway, I passed all the examinations at The University of Leningrad and began seriously listening to the lectures of Pavlov and his associates. Pavlov was almost the only person in Russia who could speak his own mind because Lenin, who was alive at that time, was practically in love with Pavlov because he was a materialistic physiologist which was eminently consistent with the Marxist philosophy of Lenin. Pavlov really gave a scientific glow to Marxism. So Lenin ordered "Never touch Pavlov; let him say whatever he wishes." He felt it was important to have Pavlov alive and well, not to have him imprisoned or destroyed. This relationship is of interest because when Lenin died, the Ministry of Public Instruction sent a circular letter to all the universities telling all the professors to devote the first fifteen minutes of their next lecture to eulogize the departed Chief of State. So everyone wondered what Pavlov would say without the protection of Lenin. On the day of his lecture the room was extremely crowded. Pavlov came in exactly at 9:00; in fact he came two minutes to nine and looked at his big watch. When the hand was straight up at nine o'clock he entered the room. He had developed this pattern of punctuality during his studies in Germany. He sat at his desk and crossed his fingers, playing with his thumbs as usual and said: "Before I continue my lecture on stomach function I have a sad thing to tell you. A great man died recently. He was a good scientist." When he said that, everyone thought he had sold out by pretending that Lenin was a scientist. After 14 minutes of eulogizing this scientist whom everyone thought was Lenin, he said, "Of course, I am talking about Professor Tegerstedt, who recently died in Europe." Everyone in the audience began to breathe freely again.

Pavlov was a sublime lecturer yet he spoke in a simple, lay, almost peasant language. He had the policy of having his assistants carrying out experiments behind his back when he lectured. The experiments, of course, were related to the subject under discussion.

I was lucky to have been exposed (weekly) to a man of Pavlov's caliber at the impressionable age of 18 years. At that time he was 73–74 years old, but his intellectual and combative vigor were unabated. Although known the world over for his "conditional reflexes," he started this chapter of his life at more than 50 years of age, when he had already reached the peak of his scientific fame for his studies of digestive glands, summarized in the book entitled: "The Work of the Digestive Glands" (1897). In 1904 (the year I was born) he was awarded the Nobel Prize in medicine. It was the first one given to a physiologist and the first one I believe given to a Russian prior to Pasternak for his "Doctor Zhivago." Even Tolstoy and Tschekov were not so distinguished.

The reason for this overwhelming success was that Pavlov introduced in this difficult field a new methodology of chronic experimentation, with very long term quantitative evaluation of the activity of different digestive glands: the salivary, stomach glands and fluids, intestinal glands and the pancreas. He became interested in this subject while a student in Heidenhain's laboratory in Germany. In order to study gastric secretion, Heidenhain made a separate pouch in the stomach of a dog and moved this pouch outside of the body so that he could study the effects of digestion in the chronic animal. However, Heidenhain's operative technique was deficient. In particular, he cut all the nerves supplying the stomach, obviously neglecting the possibility of their participation in the regulation of secretion. Young Pavlov was already an excellent experimental surgeon and in the "Pavlov Pouch" all the major nerves were preserved. Other secretions were investigated in the chronic animal by exteriorizing a "fistulae," the windows over the glandular activity during digestion. Pavlov envisaged the digestive system as a factory having many compartments in which different reagents were used for processing foodstuffs. Pavlov had to be able to study these reagents as they were secreted in response to different kinds of food in pure form. He also had to study changes observed in these reagents as a function of the type of ingested food. Pavlov came to the conclusion that the work of different digestive glands was under the strict supervision and influence of nerves. Thus he developed an impressive theory of "nervism" as he called it. In particular he claimed that following the presence of food stuff in the intestines the pancreas was "invited" to secrete its juices by activation of the vagus. Finally he could find that the secretion of the digestive glands in general, the salivary glands in particular, was of a different kind, depending on the type of food offered to the dog and that just by looking at different foodstuffs the secretion was of a different kind. He called the secretion initiated by looking or smelling food prior to being in contact with the buccal mucosa a "psychic secretion." For example, if the dog was looking at dry bread, the secretion was watery; if he looked at juicy meat it was rather mucous.

In order to get these results, Pavlov developed not only personal surgical procedures but also built a special surgical amphitheater to limit the danger of infection. He also trained dogs to be immobile when he was getting samples of these "juices" and thus could obtain consistent qualitative and quantitative results.

Following the publication of his book on digestive juices he received a piece of unexpected news. He was told that Bayliss and Starling, two London physiologists, had discovered the hormone, secretin. The injection of this hormone liberated pancreatic solution in the intestine (1902). This, of course, was challenging to Pavlov's claim that pancreatic secretin was due to stimulation of the vagus nerve. When someone in his own laboratory confirmed Bayliss' discovery, Pavlov retired to his study for half an hour. Then he returned, saying, "Well, we do not have an exclusive patent on truth." Later, Anrep (1912) convinced Bayliss and Starling, in their own laboratory, that vagal stimulation did elicit pancreatic secretion. Thus the notion of neuro-secretory regulation of body activities was born.

However, after this incident Pavlov decided not to return to the study of digestive glands "per se" but to submit "psychic secretion" to physiological analysis. And this he did until the last day of his life in 1936.

The first year of his newer studies was marred by a conflict with one of his most helpful assistants. When the assistant tried to explain the results of this experimentation on "psychic salivation" by referring to the dog's thinking, hoping, visualizing and remembering, Pavlov rebelled. He wanted to remain an objective experimenter, a true physiologist, and refused to presume that dogs as well as man might have a subjective life. And so he parted with his collaborator, who was appalled by the denigration of "man's best friend" to a mere reflex machine.

It would be incorrect, however, to say that Pavlov denied any possible subjective feelings in dogs. His view was much more restrictive. He asked whether, if we forget about anything subjective in dogs' mentality; could we understand their behavior only on the basis of objective observations related to states of cortical excitation and inhibition. Yet standing on this restrictive ground Pavlov discovered a number of facts which otherwise would not be known.

For example, if one fed a dog meat after a particular auditory signal such as 1000 c/sec tone and forced acid liquid into its mouth following a 2000 c/sec tone, one could demonstrate the following: the saliva of the dog at rest following a 1000 c/sec tone would have a completely different chemical content than after a 2000 c/sec tone. The composition of the saliva could be changed by a conditional signal without the dog actually seeing the food. The most unexpected finding was related to conditioning inhibition. The dog was made to listen to a 1000 c/sec tone prior to a subsequent feeding. Yet no food was given for a number of days despite the conditioned stimulus. Pavlov demonstrated that the usual resting salivation of the dog would drastically decrease or be inhibited.

Many people are not aware that among Pavlov's discoveries, that of experimental neurosis is perhaps the most revealing. He identified its cause with remarkable insight. His original observations were related to what happened to his dogs during the St. Petersburg flood. Many dogs locked in their cages went berserk. As Pavlov kept thinking about it he came to the conclusion that the cause of neurosis was an unresolved conflict. So he devised an experiment to test this idea. He reasoned as follows: Suppose I use a sound of 2000 c/sec as a conditioning signal for feeding and a sound of 1000 c/sec as the conditioning signal for forcing acid liquid into the dog's mouth. Thus each time the dog would hear a sound of 2000 c/sec he would have a pleasant expectation of food and each time he would hear the sound of 1000 c/sec he would expect to swallow an undesirable liquid. Suppose now that instead of 2000 c/sec the feeding signal will be 1600 c/sec and the 1000/sec signal for acid will become 1400/sec. Then surely a sound of 1500/sec would elicit a mixed emotion in the animal, who would not know what to expect.

With this technique, he found that the dog would break down due to unresolved conflict created by uncertain signals.

Let me digress to some reminiscences. In my long and busy life I have met a number of outstanding scientists. The one who undoubtedly was the greatest, Pavlov, was not however the most "brilliant." There is a difference between thinkers who by constant meditation, checked by patient experimentation discover remarkable findings such as those of Pavlov, and those who burst forth with sparkling ideas, but leave only limited achievements. Nicholai Bernstein was one of such men. So was Weiss, whose book I helped to publish, who contributed a few important ideas to rehabilitation surgery. Grey Walter, by the quickness of his mind and the brilliance of some of his ideas, was different, but his contribution was also restricted to frequency analysis of EEG and the study of expectancy waves. Warren McCulloch was another who sparkled with ideas, yet left a limited amount of solid new knowledge. Lapicque was brilliant but got himself into trouble unnecessarily. Beritoff was undoubtedly the most productive of all my personal friends yet more modest in personal conversation. He was an extremely knowledgeable and skillful neurophysiologist, having mastered the physiology of the spinal cord in frogs. He also wrote a voluminous textbook on neurophysiology and spent some time with Magoun, the outstanding American neurophysiologist. However, his most important scientific achievement was related to conditioned reflexes. He was one of the few Russian neurophysiologists who dared to challenge one of Pavlov's most crucial points.

This conflict between physiology and meaningful behavior led to Pavlov's criticism of leading psychologists; thus he unjustly scorned Lashley, the most prestigious American psychologist at that time. It is paradoxical that Pavlov, who was the instigator of about two thousand studies of conditioning by psychologists, scathed them with invectives. "Psychology is not a science," he told us, his graduate students, as often as he could.

And so he had to create his own language for this new science. An innate reflex, for example, salivation when food or acid solution came into contact with the muccal membranes was called an Unconditioned Reflex (UR). The stimulus that elicited it, the food or acid, was called the Unconditioned Stimulus (US). A neutral stimulus that regularly preceded the Unconditioned Stimulus was called the Conditioned Stimulus (CS). The conditioned stimulus after a number of associations with the unconditioned stimulus elicited a Conditioned Reflex (CR). If a conditioned stimulus was preceded by another CS, the latter became a CS of the second order, provided that it could elicit a CR. Such conditioned reflexes were called conditioned reflexes of the second order.

Conditioned reflexes could be elicited only if the CS either preceded the US, or overlapped it in time. Practically never did the CR appear if the CS followed the unconditioned one. If there was a significant delay between the CS and a conditioned response, one spoke of a delayed CR. Some dogs manifested delayed CRs after 30 minutes. If an US was repeated frequently, say every two minutes, a "conditioned reflex to time" would appear prior to the appearance of the next US. A CR to food was called a positive reflex. A CR elicited by an associated stimulus preceding a squirt of acid into the mouth of a dog was called a "defensive reflex."

Feeding the animal, or introducing an irritant into its mouth was called "reinforcement." The American terminology for this was reward or punishment. Conditioned reflexes and stimuli could be weak or strong, according to the facility with which they could elicit "*conditioned responses.*" Conditioned reflexes could be suppressed (inhibited) by different procedures. "External inhibition" was elicited when an unexpected stimulus attracted the dog's attention. Such a stimulus would elicit in Pavlov's terminology an orienting reflex, or reflex "What is it." The animal would turn its head, eyes and ears toward the stimulus. The process of attention could be studied by eliciting the "investigating reflex." Sokoloff devoted an entire session to the study of these reflexes and the related process of attention. "Internal inhibition" was produced under different circumstances. Let us repeat a CS without reinforcement and the CR will subside. Pavlov proved that its influence was due to an active inhibition, and extinction, ready to come to life at the first opportunity. Let us present another CS just before the one that we shall extinguish subsequently. This new CS will become a conditioned inhibitor. In other words, if presented before the presentation of an active CS for another CR, it will inhibit the new CR. Let us delay the reinforcement for an unusually long time and this CS then becomes a conditional inhibitor in its own right. Following an external inhibition the dog generally became drowsy, suggesting to Pavlov that sleep is general inhibition. Let us present 2 CS, one before reinforcement and the other slightly afterward. The latter will become an inhibitor in the process of "differentiation."

Pavlov enunciated several laws of conditioning. Let us create a CR to an auditory stimulus. Another auditory stimulus may become a CS also due to a process of "generalization." If one creates a CR by skin stimulation at a certain point, then a larger area when stimulated will also elicit a CR. This process is called "radiation." If one wishes to limit the effectiveness to only the original spot, one initiates a process of "condensation" by the active process of differentiation.

Obviously conditioning is due to a formation of a temporary connection between the "center" of the conditioned stimulus and the structures responsible for the UR. But is it a connection with the center of an US or an UR? According to Pavlov the former is correct. We will see that for Beritoff this was a crucial problem, one that led him to challenge some of Pavlov's basic concepts. Pavlov discovered "experimental neurosis" by presenting the dog with difficult problems to solve. Moreover, he also demonstrated that dogs of different personality types have different neurotic reactions, either hyperirritable or predominantly inhibited. Thus, as a result of compulsive thinking, day and night, Pavlov elaborated a comprehensive system of conditioning. It was impossible not to admire this man.

However, we young students had become somewhat leery about the whole story. For all, there was the initial suspicion that the dogs could "outsmart" the Master. After all, what is unusual for a clever dog to salivate when it sees food? On the other hand, to completely ignore the dog's emotions seemed to be belittling the soul of the greatest of man's friends. Also, the choice of a dog for the operative procedure was not endorsed. Pavlov used to tell us that he could not watch his wife killing a chicken, but he had no such feeling for dogs serving the higher purposes of science. We readily accepted the importance of this research for determining the limits of perceptual capabilities of the animal so obviously revealed by the methodology of conditioning. We also had to admit that the discovery of *internal inhibition* went beyond the advances of the association philosophers who also demonstrated the importance of temporal contiguity and of repetition of association for the successful outcome of associational processes. But we were not ready to concede that one of the major goals of science, the elucidation of the nature of subjective experiences such as images and thoughts, as well as hallucinations and delusions in mental patients, should be slighted.

Again the most convincing data was related to the unconscious processes of the animal, such as the composition of the "psychic" saliva when the animal looked at dry bread, or juicy meat. We had to admit that in this, as well as in conditioned reflexes established in the viscera, the CR methodology constituted a momentous discovery.

We did not like Pavlov's gross subdivision of the brain into subcortex—the site of unconditioned reflexes, and the cortex—the organ of conditional reflexes. We approved of much newer "physiological" differentiation of the brain. Despite Pavlov's denials, his pronouncements resembled the "black boxes" of other psychologists.

Pavlov used to say that a scientist must think about his experiments obsessively, every minute of the day, also at night, in order to advance his knowledge. This precept, together with the injunction of Claude Bernard (have your ears and your eyes all around your head) became the principles of my scientific life.

I believe that his most important discovery was internal inhibition resulting from non-reinforcement of a conditioned stimulus. The reason for this importance is not generally recognized, but internal inhibition is a true physiological mechanism for the retrieval of unconscious interactive processes. The lack of reinforcement gradually eliminates the conditioned reflex to the point of "extinction" (in the sense that the reflex no longer plays an active part in the immediate intercommunication processes). However, if the reinforcement attribute is re-instigated the original conditioned reflex is re-activated rather rapidly—and at times explosively! This process represents a rational mechanism for conscious-unconscious interplay.

The other outstanding contribution he made to behavioral neurology was the remarkable demonstration that if conditioning signals were too similar there was a conflict that rendered the subject erratic in its response pattern that had the character of psychoneurotic behavior in man. This type of conflict is a specific expression for the generalization of commensuratism where if two systems are closely similar, there are strong, sometimes disruptive, forces operational in the interactive process. In physical systems diffraction is a prime example; in organismal systems the interaction observed in intra-species behavior is great when compared with inter-species interaction.

We shall refer again and again to the saga of conditioned reflexes and experimental neuroses. There is no doubt in my mind that this work of Pavlov contributed more than anything else Russian to my developing scientific life. Yet I could not blindly accept his pronouncements, no matter how greatly I admired him as a teacher. I felt particularly sorry for him when, several years later, he came to Paris to lecture and to meet many of his old friends then in exile. After the lecture, Lapicque, who contributed a great deal to the mathematical laws of the process of excitation in peripheral nerve fibers as a splendid achievement of Western science, stood up and asked Pavlov what he thought about the development of "mathematical neurophysiology," and what applications he could see for it in the field of conditioning. The question was duly translated into Russian for Pavlov, who said: "Mathematics? In Physiology? Never heard about it." Everyone in the audience was sorry for both of them, for Pavlov, for his answer and for Lapicque for asking the aging master for an answer that he was so much more competent to give. As a matter of fact, such mathematical theories of conditioning based on probability of occurrence of a CR to a CS have recently been formulated.

My immediate teacher, Uchtomski, was an extremely thoughtful person. Uchtomsky became known among Russian physiologists first, and later in the western world as the father of the notion of the "dominante." Operationally, the dominante is a state of the central nervous system into which a "hyperexcited" group of neurons attract nerve potentials circulating in the nervous system, and reinforce it by their contribution. Thus in his Ph.D. thesis, Uchtomsky demonstrated that if one stimulated a decorticate cat during an act of defecation, the defecation process was accelerated by these extra stimuli previously having nothing to do with this act. In another experiment he and his student, Ulland, increased the excitability of the spinal cord neurons usually involved in the "flexor" reflex by the local application of strychnine. They then found that stimuli applied to the neurons which usually elicited an "extensor reflex," were now reinforcing the flexor reflex. Obviously, the "dominante" thought behind these experiments was to explain the plasticity of the central nervous system and the formation of conditioned connections. I later learned that the real originator of this train of thought was Wedensky himself. He was able to change a functional sequence of the cortex itself by making it hyperexcitable, and thus responsive to stimuli to which it did not usually respond.

It was Uchtomsky who gave me the subject for my first experimental work concerning the differential fatiguability respectively of the endplate and the muscle in frogs. Later on I found out that this project had been given to Uchtomsky by Wedensky himself; Uchtomsky never found time to carry out the experiments. It was this work that became the subject of my future controversy with Lapicque, Professor of Physiology at the Sorbonne, in Paris.

I based my first experimental work on the clinical tradition of physiology, stemming from the work of Claude Bernard on curare. As is well known, Claude Bernard found that the South American Indians, in order to kill their enemies, immobilized them with a material (curare) with which their arrow heads were coated. The action of this topical material was elucidated in the laboratory of the College de France. Professor Bernard reached the important conclusion that the poison used did not affect the nerve. He stated that the poison operated on the motor endplate, and that this structure, when inoperable, acted as a switch which did not allow the electrical excitation to proceed from the nerve to the muscle. In an effort to confirm the notion that the same mechanism is operational in fatigue, I performed the following experiment.

I used the classical sciatic-gastrocnemius preparation of the frog. One pair of electrodes was placed on the sciatic nerve; another pair consisted of one wire on the proximal region of the muscle and one wire on the insertion tendon. In the latter pair, the initial observations were made with the proximal electrode negative. In this case the excitatory current was ascending from the positive to the negative pole; the excitability threshold was measured. Following this operation the distal electrode was made negative, the current flow was thus reversed and the excitability threshold was measured. In each instance a brief "galvanic" current stimulus was applied. The least possible current to obtain muscle contraction was used. The threshold was lower when the proximal negative electrode was stimulated (ascending current).

Because it had long been known that it was the negative pole which excited the muscle via the nerve by "depolarizing" the membranes of these structures it was assumed that the lower threshold with the ascending current stimulated the intramuscular nerve fibers. This stimulation was thus proximal to the motor endplate. In the case of the descending current (with the negative electrode over the tendon) the muscle itself was stimulated without mediation through the motor endplate. At rest, the threshold of the nerve stimulation was below that of direct muscle stimulation. Following these determinations the muscle was stimulated rhythmically through the nerve until the muscle was apparently exhausted. At the time of exhaustion the excitability threshold was equal to both ascending and descending currents.

Interpretation: First, as with curare, fatigue exhausts the motor endplate, leaving the excitability of the nerve and muscle intact. Fatigue "turns off" the switch that permits the passage of stimulation. Indeed, the muscle remains excitable by the descending current just as well as before the fatigue. The excitability by the ascending current decreases because instead of effectively stimulating the muscle via the nerve after exhaustion of the motor plates by fatigue, one stimulates the muscles tissues by bypassing the motor endplates.

A towering figure in Leningrad at that time was Professor Bechterev. He was a great neurologist, a great psychiatrist, and wrote sophisticated books on the subject of neuroatomy. He discovered several important pathways in the brain such as the central tegmental tract; his name is associated with several diseases of neurological importance. At one time he decided to compete with Pavlov. He found in his past writings that he might be credited with the first description of conditioned reflexes which he called associative reflexes. His technique was different from that of Pavlov; Pavlov studied animals, measuring salivation, etc., whereas Bechterev observed conditioned leg movements in man, to the expectation of electric shocks.

A young neurophysiologist of repute who was working in Moscow at that time, and who influenced profoundly my future scientific development was Nicholai Bernstein. He was a physician, neurologist and mathematician of great stature. Like many of us he felt the time had come to test the sweeping hypothesis of neurophysiology not only on the legs and spinal cords of the frog and decerebrate cats, and the salivary glands of the dog, but on man. He considered integrated behavior as having the aspect of both automatism and voluntary effort. He studied locomotion in man (simple walking or running) and skilled movements such as the use of a hammer or saw at one end of the scale and piano playing by virtuosos on the other. The reason these movements escaped physiological analysis was because these actions were very difficult to analyse with precision. Several investigators in the past century, starting with Maret in France, Fischer and Weber in Germany, and Bainbridge in the United States, developed techniques for the recording of movements. The simplest technique consisted of placing small electrical bulbs over different joints; the bulbs were lighted intermittently at a constant frequency of 20 or 30 per second, the moving segments of the body, or the whole locomotion, was photographed on the still plate of a camera. Thus a silhouette of the leg or arm appeared as an interrupted line at the beginning of the movement and then a series of lines as the movement progressed. The distance between two consecutive points could be measured at the level of each joint, and therefore determine the speed of movement at that particular level. Knowing the speed, one could determine the acceleration; and knowing the mass and the moments of inertia one was able to mechanically characterize the movement in space and time with some precision. Bernstein added greater precision to the method by using 60–100 points instead of 20 points per second. In order to study the movement in three dimensions he used either several cameras, or mirrors placed at appropriate angles. With unusual tenacity, ingenuity and imagination, and having reliable collaborators, he described in detail movements of a hammer in the hands of a skilled worker. He was struck by the periodicity of these movements. Each consecutive motion followed about the same trajectory, but not exactly the same. He believed that these trajectory differences were due to feedback information which the "centers" received from the periphery in order to correct the movement. Thus Bernstein became the first neurophysiological investigator to clearly formulate the principles of feedback, and related phenomena. He believed that this kind of mechanical description of movements would permit one to penetrate the principles of nervous system activities more profoundly than those used by Pavlov. As will be seen later, I simplified Bernstein's methodology for the recording of complex movement patterns in man.

But beyond all these people to whom we were directly exposed there was always the shadow of the great Sechenov, who had now been dead for a generation.

Russia   31

In Leningrad I lived in two rooms at the top of an apartment on Fantanca, a tributary of the Neva River where I had a wonderful view of the canal. Behind the quay bordering the Neva there was a row of imperial palaces and museums. In order to supplement the meager allowance which was all my father could send me, I had a part-time job in one of the important government institutions concerning the five-year plan. My job was that of a statistical clerk, but the work was done in one of the palatial buildings from which I could see a panoramic view of the river. During the winter I walked to the University, as did many of the others, over the frozen river, a fatiguing but exciting promenade. During the time of "white nights," when the sun never set, I and my friends would spend sleepless hours breathing the romantic air of the city. In the spring, we took excursions in boats along the canals and river. The other tenants in the building were intellectuals, teachers and writers, with whom I spent long hours in discussion. I loved Russian poetry, the eternal Pushkin and Alexandre Block. Of course I was resolutely opposed to the dictatorial regime but was absorbed in my scientific endeavors, enjoying the opera and ballets, reciting poetry, discussing philosophy with my neighbors and friends and became less sensitive to the iniquities of the time.

One day however, reality unmasked itself with an unexpected accent of emergency. One of my cousins told me that the political police had come to the apartment of my aunt looking for me. Just as in Moscow, it had taken a long time to catch up with my travels. I would not say it was because of the ineffectiveness of the police but rather because of the small threat that I presented to the regime and in all probability my case was handled by very secondary people. I went into hiding again and contacted my family in Moscow. They in turn, contacted one of my uncles, another brother of my mother, who lived in independent Latvia, in Riga. He was a man of great heart and enjoyed comfortable prosperity. He very quickly contacted an organization which smuggled men and materials through the frontier. I was told to go to Poltosk, a frontier city, and wait for the smugglers who would lead me to a boarding house on the border. When I arrived at the boarding house I found another young man who was waiting for the same smugglers, with whom I had to share the room. We took our meals at the bedside, looking with envy, through the windows over the street at couples strolling hand in hand, enjoying exciting evenings, bathed in the moonlight, under clement skies. My patience grew thinner and thinner, and when no one showed at the end of the month I lost my temper, bought a ticket and went back to the apartment of my parents in Moscow. Of course they were very pleased to see me but they showed me a telegram they had just received that morning saying that the contraband people had just left the other side. So without further ado I retraced my journey back to Poltosk. The very next day, in the morning, still barely dressed, we heard shouts calling "Liberson, Liberson come down." For more than a month I had carefully avoided any contact with other human beings, other than my room mate and the maid, and now anyone cold hear my name and see my face. When we went out we found a peasant with a horse and buggy who explained to us that we had to follow him across this small town by foot, and then in the suburbs join him. We started our journey when to our horror we saw a policeman walking in the same direction. The policeman asked our driver to give him a ride. Of course the driver could not refuse that service to an officer; we wondered what we should do. At this time our driver made a sign with his hand which suggested that we had to speed up and get to his level. When we did he asked us whether he could give us a ride. We obviously agreed but without any great enthusiasm as we did not know what to say to the policeman. The latter asked where we were going and I told him a story prepared in advance, that we were asked by our relatives to put flowers on the grave of a grand-aunt who was buried just at the frontier. I doubt that the policeman believed our story, but he decided to wait and see. As we were passing a pub, the driver asked us, including the policeman, whether we would like to go and drink some vodka. The policeman readily agreed. Then we witnessed one of the most memorable scenes of my life. The policeman and the driver were drinking an inordinate number of vodkas, and getting increasingly intoxicated. After a month of impatient waiting, after having at last perceived light at the end of the tunnel, we seemed forced to forego all our plans and witness the tragic ending of our adventure at the bottom of a drunken party. I pulled the sleeve of the driver and asked him why he was destroying us. However he winked to me, and told me in no uncertain terms that I was stupid; he rapidly explained that this was a stratagem to get the policeman drunk. In fact as soon as the policeman was lying helpless on the floor we left the pub, resumed our respective places in the buggy (minus the policeman) and set out toward the frontier.

It was not, alas, the end. Soon afterwards we noticed that another carriage was following us. Obviously we reasoned that the policeman had been found on the floor and that the police were after us. We discussed the situation with the driver and decided there was no use attempting to escape, and that the best thing was to surrender with dignity. And so we stopped. To our great amazement; the party following us, stopped also. We decided that they were debating whether to be exposed to our shooting, or to shoot us first. We decided to move on. As soon as we moved, they moved too. So we stopped again. As soon as we stopped, they also stopped. Then a refreshing thought came to our minds; they were probably in exactly the same situation that we were. So we resumed our flight and came to a village where we were really stopped by young members of the Komsomal (Communist Youth Organization) with revolvers in hand. They were accustomed to see people attempting to pass the frontier and asked us for our identification cards. I then tried one last heroic trick for survival. I produced my certificate from the place of my employment which had the very imposing name of Statistical Commission. The young boy who stopped us did not know the meaning of statistics. I then put him to shame saying that he, a member of the Communist youth, should certainly know about statistics; and so convincing was my oratory that he let us go.

We reached a narrow river separating the two countries, went on foot into the shallow waters, and a few minutes later were in Latvia. I had a feeling of intense liberation, the taste of liberty at last. It was a little premature, however. The local farmers took us to a farmhouse and told us in no uncertain tones that if we did not give them everything we still possessed in terms of money or valuables they would force us back to Russia. We obviously complied with the demands. A few hours later I saw my uncle, who brought me to his home in Riga. My uncle and aunt were generous people and very understanding. Their children, my cousins, were most hospitable. It took me only a few days to fall in love with one of my cousins, but unfortunately I soon found out she was engaged. I wrote to my relatives and friends in Russia to please ask Professor Uchtomsky what he would advise me to do to further my studies. I was made to understand that he was very disappointed to learn that I had left the country, but I am sure that he underrated the compelling reasons for my flight. I received a telegram from him that I should join the laboratory of Paglioni in Rome, or that of Lapicque in Paris. In an additional letter from him, I was told that Paglioni devoted his research to the brain, and that Lapicque was "physiological," studying the excitability of nerves and muscles (an engineer and mathematician).

Inasmuch as I had an uncle in Paris who was engaged by the celebrated dancing group of Diaghilev, I eventually chose Paris.* But prior

---

*See "I Sang For Diaghilev," by Joseph Gale, *Dance Horizons*, 1982.

to this I went from Riga to Berlin for a few days, where I had my first steak with butter.

CHAPTER 2

Paris

*"Of his rhymes he was not ashamed
Cultivating solemn feelings
But in his mind clear and untamed
He questioned all fuzzy meanings"*

IT WAS IN THE FALL of 1924, after crossing by foot the frontier between Russia and Latvia and resting at my uncle's house in Riga that I came to Paris at the age of 20. It was raining. I had heard about the Latin Quarter, the residential and academic center of French students, so I repaired to that area. With suitcases in each hand I went looking for a room. To my horror there was not a single room to let in any of the numerous small hotels. Finally in desperation in one place I asked if they could rent me a bathtub for the night. The owners looked at each other with questioning expressions and then said, "Why not?" So the first night in the "City of Light" I spent in a bathtub. The maid made up my "bed" and I slept. The next morning she brought me my breakfast and placed it precariously on the edge of the tub. On getting up I moved with the feeling of lumbago from the bathtub sojourn, and my breakfast, and more importantly, the fancy teacup and the rest of the dishes went down to the floor in pieces; I escaped unhurt.

As a result of this incident I immediately went to the residence of my uncle who was in charge of a night club in the heart of Paris. He told me there was a little hotel adjacent to his club and that I could rent a room there which I did. The trouble was that it cost me about three-quarters of my budget, and that every morning I had to take the metro and ride several kilometers to the university. Incidentally, one day when I had a severe headache and returned to my room in the middle of the day I found it occupied by a young couple. Apparently it was rented every day! Because of this I again looked for accommodations in the Latin Quarter, where I now found a little room near the Sorbonne.

I soon found a way of introducing myself to Lapicque, accompanied by an interpreter. I told Professor Lapicque about my studies in the laboratories of Utchtomsky, my following the lectures of Pavlov, my research on fatigue, my political difficulties. He listened to me with sympathy and understanding. He told me that my work concerning excitability of the end plate region near the tendon could be easily explained in the light of his theories on chronaxie. He called his senior assistant, Professor Henri Laugier, and asked him to give me a place in the laboratory and to provide me with equipment. He invited me to attend his conferences; he was most encouraging and hospitable to me, a young novice of 20 years.

I returned to my room and took stock of my knowledge, and re-examined my plans. It was clear in my mind that the goal of my life was to understand the thought and emotional processes of man in physiological terms. We students of the University of Leningrad who were graduates of the High Schools at the time of the revolution were rather poorly grounded in the basic sciences of physics, mathematics, general and physical chemistry, all of which constituted the foundation of physiology of the Western Universities. My knowledge of the anatomy of the nervous system, particularly the brain, was very sketchy. But I did know that a bundle of nerve fibers, the axons, were each covered by a sheath of myelin, that insulated to a degree the electrical impulses which constituted nerve messages to the higher centers, and brain and that this sheath was segmentally interrupted by regions with essentially no myelin, the nodes of Ranvier. I knew that the myelin sheath was a part of the Schwann cell and that the nerve fiber was surrounded by a semipermeable membrane which at rest was positively polarized in relation to the interior of the axon. This polarization was due to the distribution of the ions of potassium, sodium, chlorides and calcium. Each time an electrical stimulus was applied to the membrane, depolarization occurred; successive depolarization traveling along the axon created a wave of excitation in both directions and the message was transmitted to the muscle via the motor endplate. This endplate had qualities which resembled the synapses in the spinal cord and elsewhere in the central nervous system. I knew that sensory nerve fibers had about the same structure as the motor nerves and that they transmitted the signals from the skin and muscles to the spinal ganglia, thence to the dorsal horns of the spinal cord, thalamus and to different areas of the cortex; or be directly transmitted to the anterior horn cells producing short circuit reflexes. According to the character of the messages received by the cortex and the context of previous experience, conditioning in particular, the cortex will reply to these messages by sending impulses directly toward the spinal cord, or indirectly to the basal ganglia in order to produce appropriate muscle action. I knew that in this process the central nervous system was the site of the phenomenon of excitation, of inhibition, of spatial and temporal summation, of facilitation and was responsible for conditioned reflexes. By dominant and "temporal connections" I understood that in the performance of the motor act, other structures of the brain, the rhinencephalon and the cerebellum, had to play an appropriate supportive role and that fundamentally the same patterns of function were subserved by the cranial nerves. I learned that electrical stimulation applied to the nerve in order to study the chain of these phenomena could be effective only if appropriate current was applied and that a prolonged "direct" current was exciting only at its onset at the negative pole and was only weakly exciting at the positive pole and that it did not excite during the continuous flow of current through the nerve because of the process of accommodation. I also found out that what we were taught at the University of Leningrad and what constituted the dogma of classic neurophysiology, namely that the nerve could not be excited unless there was a rapid change in the current applied to the nerve, was incorrect.

At any rate, we students in Leningrad had to start with the receptors of the skin, the exteroceptors; those which transmit the feeling of touch, pressure, heat, cold and pain. Then we learned about the proprioceptors, located in the muscles, the spindles, the golgi tendon organs; the complexity of which were unknown at that time. Then we knew that there were vestibular receptors which permitted the central nervous system to be informed as to the orientation of the head in space. We also knew about the interoceptors which were located in the stomach, in the heart, in the vessels, in all viscera. And finally we were made aware of the "distant" receptors which responded to photic stimulation of the retina, and the cochlear (auditory) and olfactory receptors. Our teachers stressed the importance of the distal receptors for alerting the organism to situations of potential embarrassment, including danger, that allowed time for preventive maneuvers.

The sensory nerve fibers travel toward the spinal ganglia which contain bipolar cells that not only transmit the impulse, but also nourish the fibers, since if the fibers become separated from the bipolar cell they degenerate. The bipolar cell sends a proximal branch toward the posterior horn of the spinal cord. In the case of a simple reflex, the monosynaptic reflex, it will make a junction with the anterior horn cell. The latter constitutes a central station sending nerve impulses into the ventral roots toward the plexuses, and peripheral nerves until it reaches the motor end plate and muscle; the anterior horn cells also have nutrient properties. The sensory fibers running in the spinal nerves are also directly and indirectly connected with the ascending fasciculi in the spinal cord through the spinocerebellar and spinothalamic tracts. Thus these long fiber tracts participate in the arc reflexes through the cerebellum, or thalamus and even the cortex, with or without participation of other nuclei of the brain. In the reverse, we knew that the cortex and basal ganglia, and some of the brainstem nuclei initiate descending tracts in the spinal cord to reach, directly or through internuncial neurons, into the same anterior horn cells. We knew the concept formulated by Sherrington of the "final common path." This concept expressed the idea of many descending cerebral axons, or internuncial cell axons impinging on the anterior horn cell bodies, or their appendages (dendrites) competing for a final transmission of those stimuli that will be permitted to travel in the ventral roots in order to stimulate appropriate muscle fibers. We became acquainted with the experiment of Claude Bernard in which he discovered that when the nerve trunk was stimulated the impulse traveled in two directions, downward toward the end plate (or sensory receptors) and upwards toward the synapses in the central nervous system. However, we knew that stimulation of the muscle did not result in the nerve impulse traveling from the end plate back to the spinal cord; and that stimulation of the sensory cells in the posterior horn would not produce impulses back to the sensory fibers.

We were introduced to the work of the Dutch physiologist, Magnus, who showed that the posture of the animal and the relation of the position of the head and body was regulated by a series of reflexes. One reflex leading to an event would elicit a reflex of the next order so that the entire chain beneficially modified the flow of movement during jumping and falling. We also knew about the experiments of Fritsch and Hitzig, who showed that stimulation of the animal's brain cortex had a somatopic orderly distribution of muscle movements. But most of our information concerning electric properties of tissue was of the nerve itself. We were told in a rather picturesque way about the regional experiments of Galvani, with Mrs. Galvani preparing frog legs by hanging them, before cooking, on the metal bar of their balcony. She was horrified that the legs jumped spasmodically. She called this to the attention of M. Galvani, who believed that it was the metal bar that did the stimulation. We were then told of the famous controversy between Galvani and Volta, where one maintained that the muscle movement was the result of electrical stimulation, and the other maintained that the electricity originated in the nerve itself. The solution was achieved in the experiment where two nerve muscle preparations were utilized showing that both men were correct.

Just before my departure from Russia I learned that Loewi investigated the transmission of the inhibitory action elicited by stimulation of the vagus nerve on the heart, by an elegant experiment. He collected the fluid of perfusion resulting from intense stimulation of the vagus nerve and placed the perfusate into the heart of another frog; the perfusate stopped the heart action of the recipient. He thus opened a new chapter of neurophysiology concerning chemical transmission of the nerve impulses. Chemists soon isolated the transmitted stimulating substance as acetylcholine.

Several decades before my coming to Paris a Dutch physiologist, Hoorweg, in a very brief paper showed that there should be a critical minimum time during which the current must flow through the nerve before excitation ensued. Indeed depolarization, which is the basis of excitation, requires a specific time to be effective. Shortly after Hoorweg's paper a French physiologist, Weiss, showed in a very convincing experiment that there was a definite relationship between the intensity of the applied current and the time necessary to achieve depolarization. Weiss's experimental set-up consisted of two wires separated by a variable distance in front of a gun. The bullet projected from the gun would first sever the closest wire and then a finite time later the second wire. The system was so arranged that breaking the first wire closed the electrical circuit that generated the stimulus to the nerve in the muscle; breaking the second wire turned the stimulus off. The curve which showed the relationship between the duration and its intensity was designated as the "strength-duration curve." The shape of this curve may be easily deduced from the development of the processes of depolarization that takes place during the stimulation (exponential). Lapicque confirmed the work of Hoorweg and Weiss, and showed that there were some consistent experimental deviations from the mathematical models. But Lapicque made a daring generalization of the mathematical theory to be applicable to all excitable tissues.

According to the theory of Lapicque all excitable tissues, nerve and muscle, obey the law of Weiss, expressed by the strength-duration curve. However, the scale of time is not the same for all tissues. Thus for the sciatic nerve, the scale of time is in fractions of a second; for a muscle of the toe the time scale is longer, and indeed in the case of the stomach the time scale must be in seconds for superimposition of the strength-duration curves. And so the concept of Lapicque was that one should express all the processes of excitation and inhibition of the nervous system and muscles of the various animals by the same general law if the time courses are in proper units. He called this physiological unit "chronaxies." He demonstrated experimentally that the chronaxie could be determined by doubling the intensity of current (of indefinitely long duration) required for stimulating a tissue. He called this intensity the "rheobase." He then postulated that the nerve impulse travels a meter in a chronaxie no matter which tissue one considers. Thus he passed from the consideration of time of excitation of nerve to the consideration of time in conducting all messages.

There appears to be a common tendency of scientists to postulate a simple, general law designed to explain a great number of phenomena which results in an enslavement of their minds. This theory was quite attractive in its simplicity. But I soon learned that simple theory does not work when applied to complex processes as those of the nervous system. Thus like Wedensky with his parabiosis, like Uchtomsky with his dominanta, like Pavlov with his conditional reflexes, Lapicque's theory of common physiological time (chronaxie) was far from explaining "all" of the functions of the nervous system. And yet, just as in the case of Wedensky, Uchtomsky and Pavlov, the formulation of such theories are never without partial significance. We will see later, that in other forms the ideas of the importance of time factors in the processes of excitation, inhibition and particularly of conditioning, largely ignored by the classics of neurophysiology, were in some cases of crucial importance. These time factors were brought to the attention of the scientific world by Lapicque, but because of his dogmatism and because of his impossible claims, this brilliant investigator lost credibility. Nevertheless chronaxie did find a considerable useful application in early neurological electrodiagnosis and therapy. It is the unfortunate of great minds to try to dogmatically generalize the value of their discoveries which, in the final analysis, undermines their reputation. If for example Wedensky just stated that plunging a segment of the nerve into narcotizing fluid is an interesting experimental device to study certain characteristics of excitability of the nerve, instead of saying that this is the model of all excitatory and inhibitory processes in the central nervous system, his name might have been much more generally known. If Uchtomsky had stated that among the infinite variety of relationships between the concurrent stimuli one interesting phenomenon may be described under the name of "dominanta," instead of saying that dominanta may explain all the results of Wedensky and Pavlov, his name would have traveled beyond the frontiers of Russia. If Pavlov, instead of stating that conditioning is the only process which permits animals and man to learn and conceptualize, and that the understanding of conditioning is the key to understanding mental disease, would have simply stated that certain manifestations of behavior follow the rules of establishing conditioning reflexes, and conditioned reactions, his name would be less controversial than it is at present. Further, if Lapicque instead of dogmatically stating that all nerve processes comprising the central nervous system of all animals on the earth, and in the seas are found to be identical, if their durations were measured in chronaxies, had stated that there is an important differentiation between various tissues, according to the duration of the similar excitatory, inhibitory and conductive processes his name would have been free of the sarcasm of which it became subject to. Finally, if Sherrington had said that in certain cases, inhibition is associated with excitation in the antagonistic systems, instead of saying that all inhibitory processes are of a reciprocal nature, ignoring the discovery of Sechenov, of generalized inhibition, now known under the name of "widespread inhibition," his already great reputation would have been enhanced.

These great neurophysiologists would forever be remembered as wise and decisive pioneers of the knowledge related to the developmental studies of animal and human behavior had they not been so inclusive in their statements.

Yet, as it stands now the memory of Wedensky is obscured by the statement of many physiologists pointing out the inherent absurdity of some of his theories. The contributions of Uchtomsky are practically unknown in western science. The anti-Pavlovian statements are abundant in the psychological literature. And the name of Lapicque is castigated because of the findings which were contrary to his claims. As a corollary to his theory, Lapicque felt that in order for a nerve pulse to be transmitted from one neuron to another, or from the nerve to the muscle, these tissues should have the same chronaxie. That, inasmuch as each stimulus applied to the nerve results in a contraction of the muscle, one has to have the same chronaxie for the nerve fibers and the muscle fibers;

the role of the motor end plates was completely ignored. In order to confirm his theory Lapicque would determine the chronaxie of the sciatic nerve and then that of the gastrocnemius muscle that was activated by the sciatic nerve stimulation. He would find the same chronaxie. However, stimulation of the muscle where the sciatic nerve penetrates the muscle is only apparent; in reality when one stimulates the muscle at this level one stimulates the intramuscular nerve fibers! My experiment performed in Leningrad was particularly suitable to test the basic theory of Lapicque. Indeed, by reversing the current flow between the two needle electrodes inserted at the two polar ends, one in the area of the end plate and the other near the tendon, I was able to test the chronaxie of each independently. And they were different.

On returning to work in the laboratory I explained to Professor Lapicque my research plan. The mere suggestion that his theory might be wrong elicited a paternal condescending smile. He said that he was confident that I would modify my views and that, after learning his technique from Professor Laugier, I would become his staunch supporter.

I was quite pleased with this turn of events. I was given a small laboratory of my own, fully equipped, and a dozen frogs. I was "on," facing my new destiny as a professional neurophysiologist. I was taken seriously! Unexpectedly the transition from a Russian University to a French one could not have been smoother.

Within a few weeks I compiled 10 experiments after being coached by Professor Laugier and his assistants, who were exceedingly accommodating. Every experiment confirmed my previous conclusions; these were not consistent with the "tuning" hypothesis of Lapicque. Fatigue seemed to effect the end plate first, and, by doing so, to interrupt the passage of the impulse from the nerve to the muscle.

Following this work I requested another interview with the Professor, whose fame had suddenly whirled up, as he had just published his book, in which he expressed the results of his detailed thinking. He raced toward the Nobel Prize competition with flying colors. Everywhere in the laboratory one could see samples of letters supporting his candidacy, to be sent to the members of the Nobel Committee. In this exciting milieu I said: "Professor, I believe that you have been mistaken, your theory is untenable. Each of my experiments speak for themselves; there is no tuning. The end plate is the switch which connects or disconnects the nerve with its muscle, as your celebrated Claude Bernard stated in this very city, a half a century ago on the basis of his experiments with curare." As noted earlier, in order to build a tuning theory Professor Lapicque

Paris

invented a biological unit of time which he called chronaxie. Each functioning tissue counted its time in chronaxies, not in milliseconds. When the chronaxies are the same for contiguous tissues, like nerve and muscle, they become tuned, and the impulse if present, may pass from one to the other, if not the same, the impulse is blocked.

My experiments showed that before fatigue interrupts the passage of the nerve impulse, the chronaxie of the part of the muscle devoid of end plates is longer than that of the portion of the muscle containing end plates. After muscle fatigue the chronaxie of the portion containing end plates becomes *equal* to that of the purely muscular portion. In other words, my findings were in direct contradiction to the theory of Professor Lapicque.

As I was talking the Professor's natural left eye twitch quickened, but aside from this involuntary emotional response, Lapicque remained calm and paternal. He said, "I am truly sorry that you came to this devastating conclusion, my young friend. What you found must have some other explanation, since all the world famous physiologists are now rallied to my side. I just received letters from Professor Adrian and Professor Hill to this effect."

To this I replied, "I do realize the importance of my findings, sir, and to me they command the conclusions that I formulated."

There was a moment of silence and then he said, "I'll tell you what I will do. Write a little note in which you will describe your experiments, and their results. Then add a simple sentence to the effect that you do not quite understand these results as they seemingly contradict Lapicque's previous findings with curare. If you write such a note I will personally present it at the next session of the Academy of Science. Then I will give you some other research subject to investigate in my laboratory."

The blood rushed to my head and I heard an inner voice say, "Don't capitulate, fight for your ideas, and make him recognize his mistake." Furthermore, at that time I was still belligerent of the authoritarian power behavior of Stalin and therefore I could not easily compromise with Professor Lapicque in free France.

The drama of this confrontation was enhanced by the fact that Professor Lapicque was important. Prior to his lectures a clerk with a silver chain preceded the Professor to announce his coming. He was lauded by his contemporary physiologists, and here I was a novice of 20 years!

But I said, "No Professor, no. I am quite prepared to repeat my experiments as many times as you wish. I am not puzzled by my experiments as you seem to be. I believe my experiments clearly indicate that your theory is erroneous. As to your research with curare, I would be delighted to repeat your experiments, in your own laboratory, and to try to understand the difference between our results. But if I am not mistaken, it appears that you did not explore separately the two regions of muscle as I did in my experiments."

Following this repartee Professor Lapicque said, "Well then I don't think that we can continue to work together." As a final consequence I moved to the Faculty of Medicine, where I received my training in electrodiagnosis and electrotherapy, and have remained a diagnostician and therapist until this day.

As a sequel to this dramatic event one day when I was vacationing in Brittany, I received a postal card from Professor Lapicque stating that a Professor in Germany sent him a paper in which he repeated what I had done with similar conclusions. He informed the writer that "my young laboratory assistant, Liberson, had earlier done the same experiment." When Lapicque published his second book on Chronaxie he discussed in detail my work and found some explanation for my findings that did not contradict his theory. But the whole controversy was resolved when it became evident that signal transmission from nerve to muscle, and in the CNS from neuron to neuron, was chemically mediated.

At the end of my medical studies I worked in the Department of Doctor Bourguignon at the largest neurological hospital in France, the Salpetriere. Professor Bourguignon was the head of the Department of Physical Therapy and Electrodiagnosis.

During the years of medical school and training, I met and married my first wife, Mira, who also was a medical student. At that time our earning capacity was not great; but by a twist of fate, Henri Laugier, the former chief assistant of Professor Lapicque, was walking down the Boulevard Saint Michel and I met him. He stopped, was very friendly and invited me to have coffee with him. He asked me what plans I had. I confided that after the end of my Salpetriere job I had no plans, but that I dreaded becoming a general practitioner. Laugier then stated that he had recently assumed the Directorship of the Department of Work Physiology in one institution and of an Electrophysiological Clinical Laboratory at a hospital. He also stated that he believed the theoretical work of Lapicque had reached an impasse and that neurophysiologists, instead of bumping their heads against the stone wall of our present ignorance of fundamental life processes and the nature of the mind, should apply current knowledge to the practical problems of the day.

Laugier, together with a professor of experimental psychology and a psychiatrist, had established a Society of Biotypology. He explained the aim of this new movement as follows: Men and women are different from one individual to another by many characteristics; different kinds of intelligence, different types of memory, different degrees of emotionality, of resistance to the environment, different abilities, artistic inclinations, sex drives, etc., etc.. Most of these differences may be quantified by specially developed tests, so that each individual may be characterized by a psycho-physiological profile of his or her own, in which age and sex should be taken into consideration.

On the other hand, different occupations require different types of individuals and different aptitudes. In a rational society the requirement of a job should be matched with the individual's characteristics, so that vocational orientation and selection should be done on a scientific basis. And then he added:

"I have gathered around me a group of young talented psycho-physiologists and we have enthusiastically started to work. I do not have at present a full time job for you, but I can offer you part time sessions. This will grow, I am sure, into a full time research position. Would you be interested?" Would I? I would do anything to get out of the prospect of becoming a general practitioner! This offer was doubly opportune as my wife had just divulged that she was pregnant, and this would ease her worry about future finances.

Laugier had the incisive intelligence of a brilliant Frenchman, which was contaminated by practical considerations, from a Russian point of view. He was tainted by atheistic convictions and ideas of social justice that I admired. He wrote beautifully and his lectures were lucid; he wrote his letters in long-hand and maintained exhaustive files at his home. He also had a most impressive private collection of French painting. I found out that he was quite active in social-radical circles, and had powerful friends in the Government. He soon became general manager of the Foreign Affairs Ministry, and later Director of the French National Research Council and Assistant Secretary of the United Nations for Cultural Affairs.

At the psychiatric hospital which he directed I became engaged in familiar activities. In the Department of Physiology of Work I was asked to participate in some unusual research. Laugier accepted a contract from a manufacturer of shovels. I believe that this manufacturer was convinced by Laugier that the selection of shovels should be on a physiological basis. Indeed, shoveling the same material with shovels of different shapes consumed a different amount of calories, presumably less in a better shaped shovel. In other words, the energetic cost of the work with a particular type of shovel became the criteria of its efficiency. I accepted this premise, learned how to measure the energy consumption of men and completed the assignment. I remember that the famous Professor Janet, the pupil of Charcot, and contemporary of Freud, visited our laboratory and somewhat deflated our enthusiasm. Janet, well in his eighties, was skeptical that relatively small differences in energetic cost would make a significant difference in worker productivity. He said, "Pay them twice as much, and they will produce three times as much work; that is the motivation key for productivity." However, another famous visitor, the British, A. V. Hill, who contributed immensely to the physiology of muscle was encouraging. Incidentally, I showed him a Soviet translation of his book that he had never seen. He inscribed it for me: "To my astonishment." I visited him later in his laboratory and met the young Rushton, who had published his criticism of Lapicque's work, before I was allowed to do it.

The energy cost for a human or animal was determined simply by measuring the amount of oxygen consumed by the individual per minute. For these measurements the subject wore a mask containing a known mixture of air. One calculates the difference between the percentage of the inspired and expired air, and multiplies this difference by the amount of air ventilated per minute. The additional knowledge of the exhaled CO_2 permitted one to translate the values of oxygen consumed into calories. At the present time such gas analyses is done electronically and automatically, but at that time it took me a major part of my working day to carry out this analysis.

One of the most general physiological laws is based on this type of respiratory gas analysis. It states that, at rest, the minimal oxygen consumption, and therefore caloric expenditure, is proportional to the total skin area of the body, throughout the entire warm-blooded animal kingdom. It is a most intriguing law.

One of the first things that I wanted to do was to reproduce in man internal inhibition in cardiac function, therefore remaining outside of a voluntary cerebral action. I succeeded with the help of Doctor Marques, who later became a Professor of Radiology in Toulouse.

We had in our laboratory of the Conservatoire des Arts et Medicine a well functioning electrical ergometric bicycle that was ordered for research in fatigue. Its rear wheel was made of solid metal which rotated between the poles of a powerful electromagnet which provided a resistive power to impede its rotation. I designed an experiment where the subject would come to the laboratory every morning at the same time and would exercise on the bicycle for a fixed period of time. His pulse was recorded before he started to pedal, during pedaling and for four intervals after pedaling. One day we did not tell the subject that the electromagnetic was not energized and therefore the wheel turned freely. When the subject made a sign to us that something was wrong with the experiment we told him that he should pedal as usual. At the end of the usual period of the experiment we energized the magnet and then "reinforced" the experiment. We continued this routine for several days. We found remarkable effects displayed in the statistics. At first the subject showed a tachycardia which could be ascribed to the emotion due to the change in experimental conditions. However during the succeeding days, with no reinforcement, to our bewilderment and joy, with the regularity of an involuntary mechanism, the pulse slowed down below the usual resting level during the course of a week or so. Then progressively the heart rate returned to its normal level. Clearly the subject did not tell his heart to slow down; obviously this was an unconscious process. There was no doubt that we did reproduce internal inhibition of the cardiac rate due to a completely involuntary action, confirming the generality of the Pavlovian discovery of internal inhibition due to the lack of reinforcement.

We were very proud of being the first to devise such an experiment, and my credibility as a scientific worker reached an enviable level among the personnel of the laboratory.

While working in Professor Laugier's laboratory a singular happening occurred that has absorbed me, and my thinking, for much of my scientific life. A graduate Ph.D. student in psychology from the U.S., Herbert Jaspar, who was destined to play an important role in my professional development, had been studying chronaxie in man, for his dissertation, in Paris. He visited our laboratory shortly after he had visited Hans Berger in his laboratory in Jena, Germany, and reported the remarkable observation that the human brain showed spontaneous rhythmic electrical activity, and that he was planning to do research in the direction suggested by Berger. This rhythmic brain activity was influenced by various physiological happenings, such as opening and closing the eyelids, and during various stages of awareness, including sleep. Dr. Berger was a psychiatrist, but he used high quality, high gain, low frequency amplifiers supplied by the Siemen-Halske Company. He used invasive electrodes on members of his family and his psychiatric residents;

he was not loved by the latter. He first published some findings in 1929 (Uber das elektroenkephalogramm des Menschen), but it was not until Adrian and Matthews in 1934 reproduced Dr. Berger's findings and added other features, that the science of electroencephalography (EEG) was really established. Actually, Berger did not discover brain electricity. Several Russian and Polish physiologists before him reported its presence during animal physiological experiments. The most complete account was made by Pravdich Neminsky, professor of physiology at the University of Kiev, my native city. He published a paper in Pfluger's Archives, the leading physiology journal at that time. The paper was submitted before World War I. It described the presence of several waves of electricity recorded from the cortex of the dog. Some of them were relatively slow on the order of 10 per second and others about twice as fast.

Following Jasper's visit, Dr. Berger was invited to Paris and presented some of his work. Professor Laugier was deeply impressed by this new phenomenon and suggested that one of his assistants should develop an EEG program in the laboratory. Unfortunately this assistant's wife, who was his subject, had almost no EEG activity that could be recorded with the instruments then available. Discouraged, he abandoned the project and said to me: "Why don't you do it, you believe in psychiatrists." Following this, I brought my wife to the laboratory and found that her alpha waves were a "foot" high. I was launched as an electroencephalographer, with a one-channel instrument and an oscillometer of Dubois presented to me by Fessard.

Berger used electrodes screwed into the human cranial periostum and to his jubilation reproduced Neminsky's waves which he called alpha (slow) and beta (fast) waves. He then found that the presence of either of these waves correlated with the state of mind of the subject. Slower waves were present at rest (mental rest, not drowsiness) while faster waves were correlated with the mental activity of the subject.

At the end of the '30s the Nobel Prize winner in electrophysiology, Adrian, a British neurophysiologist, confirmed Berger's main findings. He also discovered some new phenomena. In a subject with an EEG characterized by a frequency of 10 c/sec alpha activity, but submitted to excitation by bright light pulses of 11 c/sec., the dominant frequency of the subject changed from 10 to 11. It was proof that the brain activity of a subject can be influenced by the environment. Several Russian and American investigators discovered that stimuli of different nature and frequency could influence the frequency of individual EEGs.

One must bear in mind that Berger's discovery was difficult to accept by neurophysiologists because of the general contempt for gross electrodes. We felt only microelectrodes could tell the real story of significant brain activity. Nevertheless I was trying to prove that auto-rhythmicity was not only present in muscles and nerves but was present in the human brain. Moreover we could ascertain that changes in the state of consciousness, mental concentration, sensory perceptions could be recognized by those who knew how to capture them. And most of all I was witnessing this marvelous phenomenon expressed by the fact that a very slight shadow of drowsiness, that imperceptible wandering of the mind during the early imagery of beginning to fall to sleep, would immediately express itself by depression of alpha activity, by the appearance of theta waves (5–7 per sec) and then by the emergence of bursts of 14 per sec activity, the beautiful spindles that were appearing on the face of the oscilloscope which I had purchased for the laboratory.

Of course, I was committed to Professor Laugier's projects of differential bioanthropology so I was compelled to see if there were individual differences in terms of the EEG between normal and abnormally behaving subjects. If differences were found, would they correlate with other psychophysiological indices. First of all, I found that a certain proportion of normals did not have alpha waves, at least with amplifications available to us. I therefore painstakingly quantified the amplitudes of the alpha waves in various individuals to establish the distribution of the amplitudes and frequencies. Then because certain subjects fell asleep almost instantly following eyelid closure, and others despite long recordings never showed sleep patterns, I subjected this attribute to a correlation study. The laboratory in which the railroad engineers were studied anthropologically, physiologically and psychometrically offered an opportunity to find correlations between the EEG and many psychophysiological features.

One day, a young collaborator of Piron and Fessard came to me and said that he had possession of a drug, mescaline which produced vivid hallucinations in normal people. In addition this drug produced a degree of excitement and depersonalization. He suggested using a subject who was a psychologist working with Dr. Piron. We ran control studies with no drugs and then during and after injection of mescaline. We found, to our surprise, that during the hallucinatory phase there was a drastic decrease in alpha frequency activity. Thus for the first time we demonstrated that psychotropic drugs producing hallucinations, depressed EEG activity.

On other individuals I studied the effect of small quantities of drinking ethyl alcohol beverages, particularly in subjects who under normal conditions did not have a stable alpha rhythm. Under the influence of alcohol such individuals tended to develop well defined alpha and theta activity. Thus in 1934 the first correlations between the EEG and neuropsychopharmacology were published.

In the laboratory work with the engineers no significant differences were found between those individuals with and without spontaneous alpha rhythms. However, in those who had alpha rhythms, those with well sustained and regular rhythms, the psychophysiological indices were more favorable to operational success than those individuals whose alpha frequency activity was irregular, or mixed with a great deal of fast activity.

We were of course all excited to find that the slightest changes in consciousness were expressed by changes in the EEG. But we were impatient to determine whether characteristic changes were associated with more sophisticated states of mental activity. We did indeed find that in subjects who performed difficult arithmetic calculations in their heads, the occipital alpha waves would disappear, and alpha frequency waves would then appear in the frontal and temporal regions. I described this phenomenon together with Laugier and was both pleased and surprised that many, many years later it was described by other investigators in America under the name of kappa waves. The scientists of France had to become accustomed to this kind of ignoring of publications in the French language. As another example, Fessard and his co-worker published a basic observation concerning the facility with which the subject could either suppress, or reinforce alpha waves at will. This was accomplished by the subject forcing himself to have visual imagery (suppression), and by pretending to listen to auditory stimuli (reinforcement). Those who were responsible, in the U.S., for the "new" game of bio-feedback by reinforcement, or suppression of one's own alpha rhythm never quoted the work of Fessard and co-worker.

Thus Fessard and his co-worker, as well as myself were the first in France who were publishing work in electroencephalography. I published several reviews which acquainted the French scientific public with this great discovery.

It is really inexplicable why Berger did not receive the Nobel Prize in Physiology for this discovery which I think he should have shared with Pravdich Neminsky who was alive at that time. All his work was fully confirmed in the laboratory of Adrian, and by others the world over. Instead, Berger was subjected to indignities during the Nazi regime because of his ancestry containing some Jewish names. He was driven by Nazi oppressors to a state of profound depression which resulted in his suicide. One of my Russian friends who finished his training in Jena, where Berger was the Professor of Psychiatry brought me a few original EEG tracings of Berger which were lost during my departure from France.

My publications on EEG attracted the interest of psychiatrists and neurologists. One day I received a visit from Professor John Delay, then a young man who was involved in preparing a dissertation on memory in the laboratory of Pieron. He came to collect information that pertained to our brain wave studies which I fully shared with him, only to find out a few months after our interview that he had published the first book on EEG in France.

In 1936 Professor Pagnez and his associate Pilchet asked permission to send their epileptic patients to me for study as at that time I was the only physician in France who could record EEGs. Dr. Fessard, the neurophysiologist at the Sorbonne lent me a pre-amplifier for my equipment that could be used for recording DC slow potentials. For the first time it was shown that 3 c/sec wave and spike potentials of Petit Mal is associated with a slow DC potential. I remember Dr. Fessard visiting us after the war and marveling at our American kitchen. He told us the story of an American Nobel Prize winner in Physiology who came to Paris, during a time when food was very scarce. The Paris physiologists saved food for a month for a sumptuous meal, but he told them he was on a strict diet and asked for a tomato sandwich.

Dr. Pagnez was the Lennox of Paris with the largest clientele of epileptics in Paris. He and his associate had a common interest in experimental Brown-Sequard epilepsy. Brown Sequard observed that if in a guinea pig one disarticulated one lower extremity at the knee joint, a few weeks later the animal developed a peculiar convulsive state which consisted of torsion spasm and uncontrolled movements each time the experimenter touched the animal's neck. Later it was demonstrated that because of the inability of the animal to scratch the ipsilateral side of his neck due to the amputation, there was a proliferation of lice in that region and the uncontrolled movements were merely an expression of frustration. When the neck region was washed, the seizures stopped.

In 1938, I recognized that the fundamental frequency of EEG activity was 3.3 per second; all other major EEG frequencies observed were simple consecutive integral multiples, such as 3.3 times 2, 3, etc. In fact this 3.3 cps frequency is the most constant rhythm exhibited by some epileptics. This arithmetic relationship discovered early in my career has been so consistent, and so vital, over the years (my most recent paper on this subject appeared in 1993) that I have equated it with "the language of the brain."

As noted earlier, Adrian in England (1934) discovered that brain waves may be "driven" by frequencies other than the alpha band, in certain individuals, by intermittent photic stimulation. I in turn, showed that after the stimulation is cessated, the brain may continue to "oscillate" at the frequency used for stimulation, as if the brain "expected" the stimulus to continue. Later W. Grey Walter showed that expectancy waves may easily be studied by an averaging methodology. Livanov, in Russia, demonstrated that if one used a flicker light as a stimulus for conditioning, the "driven rhythm" was indeed detected in the EEG recorded during the inter-test periods. This was later confirmed by John and collaborators in the U.S.A., but we observed that the latter phenomenon occurred only while the animal was being trained, as if these "induced" waves were directly related to action and not consolidated memory.

Thus in different laboratories of the world medical scientists became excited by the new possibilities offered by recording the electrical activity originating in the brain, without invading the brain. We exchanged information as to our findings in publications and by correspondence with fellow workers. We were all fascinated by our data and by the same passion to understand elements of brain function. A relatively small number of "brain wave friends," ever increasing in number, emerged in the various countries.

Despite these wondrous happenings in neurological medical science war clouds were ominous and in 1939 World War II began. Late that year the Maginot Line was breached, and in early 1940 Paris and much of France was occupied. Toulouse was not in occupied territory, so our little family migrated to that city. While in Toulouse my wife learned that Mrs. Eleanor Roosevelt was interested in the intellectual Mensheviks and was offering them asylum in the United States. Mira was transferred (as a physician) to the Catalonian area, where she lived with our young daughters. There in carrying-out her medical duties she visited her patients on mules and donkeys and endured many hardships (of which she has published some of her adventures). I repaired to Marseilles to the American Embassy for a visa to the U.S.A. There the red tape unfolded. But finally the Consul bluntly stated that it was his job to determine if I was an "intellectual." Fortunately I was able to show him a current American Annual Review of Physiology in which quite exceptionally, not one, but three of my published papers of the preceding year were quoted. He then said "I am satisfied that you are an intellectual." With the granted visa my family went to Madrid, and then to Lisbon. For various administrative reasons I could not receive a personal visa to Spain, so I was forced to go to Casablanca, Morocco, and after some months of waiting, eventually joined my family in Lisbon; from there we came to "The Promised Land" as an intact family unit.

This story had an interesting sequel. Because of the help that was given to me indirectly by the Annual Research of Neurophysiology, I was able to write a chapter on Russian neurophysiology. In this chapter I was led among my different reports to strongly condemn the Stalinist oppression of Russian scientists. Inasmuch as the Annual Review of Physiology is an indispensable book for physiologists the world over, it was not suppressed in Russia and my chapter was avidly read by all the Russian physiologists.

This had two outcomes, one a trivial one and the other much more important for my future work. My daughter Helen spent a semester at Northwestern University when a group of Russian students came to visit. Inasmuch as she spoke Russian she was asked to guide the group. "Is your name Liberson?" asked a student in physiology. She said yes. "Do you have a relative in the United States who is a physiologist?" Helen answered, "Yes, it is my father." When the student heard her answer he took out a notebook from his pocket and asked all sorts of questions, stating that I was unfair to the Soviet government, which scared my poor Helen.

The other effect of my chapter was that Professor Beritoff whose work I had praised asked me during his visit to Europe to translate his book. Not only did I do it, but I derived a great deal of information from this great physiologist.

CHAPTER 3

U.S.A.

"He dreamed he was a man of science
Who understood the non-compliance
Of certain facts that refused to bend
To a universally accepted trend"

AND SO FOR THE second time in my life I escaped almost certain death. In December 1940 my family and I saw the Statue of Liberty in the great city of New York. We experienced the most wonderful reception full of consideration and understanding. We were placed in two rooms of a rooming house and were showered with kindness and understanding, which helped compensate for the fact that we were living in cramped and straitened circumstances with a group of refugees.

My wife Mira, a radiologist, could not find a job as easily as I, because she was not so well known in neurological circles, especially since I was known in the new exciting field of EEG. And then I was the head of the family and she had much to do with the children on a very strict budget. The problem became much more serious in the ensuing months.

Unfortunately, in my rigid European ways I was not always equal to this outpouring of sympathy. Thus on Christmas Day, my two little girls proudly came to me, each with a crisp new dollar bill. "Where did you get that," I asked, fearing the worst. They said a gentleman in the front room gave them the money. Horrified, I instructed them to return the money at once with thanks and to tell the nice gentleman that your parents forbid you to take any money from strangers.

There were other blunders. I was not accustomed to such wide acceptance for Jewish people. One day, wishing to buy some ham, I went to a store which looked appropriate to me and asked for some. "We don't have it," was the answer. The next day I went to the same store and asked the same question. The exasperated salesman took me by my sleeve, dragged me out of the store and showed me the sign that I missed. It said "Kosher."

This reminded me of an early adventure in Paris when my knowledge of French was also rudimentary. Once I took a bus, sat down and continued to read my book, which was "Secreter Interieur" (Internal Secretions). Soon I was distracted by the chattering of two pretty French girls and almost missed the stop where I had to get off. When in the street, I mournfully realized that I had left the book in the bus. I took the next bus and went to the terminal. There I asked for my book. However, I forgot the fineness of the French language which includes a slight difference between a book and a pound. Ignoring this subtle distinction I told the man that I had left "une livre" in the preceding bus. The inspector lifted his brow and asked, "Une livre de qua?" I said a pound of internal secretions. He laughed and gave me my book.

I was also most cordially received by the American "brain wavers." We were a small international family and each of us knew almost precisely the work of the others. Drs. Hallowell Davis (Boston), Herbert Jasper (Iowa), Fred and Erna Gibbs (Boston), Charles Stephanson (Vermont), Hans Strauss (New York) and many others were most helpful; particularly Jasper and Strauss. With the latter, I started to collaborate almost at once. Although my papers indicated that I was to have a two year research associate position in Hartford, Connecticut at the Hartford Retreat, the setup at Hartford was not yet ready. Therefore, through the kindness of Professor Israel Wechsler, Chairman of the Department of Neurology at Mount Sinai Hospital in New York and Dr. William Bierman and his assistant Dr. Sidney Licht, I was offered a jewel position as a Fellow at Mt. Sinai Hospital. My neurological half of duty was in the EEG laboratory of Dr. Hans Strauss who became a friend until his premature death. My physical medicine duties were with Dr. Licht who spoke French and became a life long friend.

Shortly after I came to his department, he said, "Liberson, I've scheduled you to talk at the Academy of Medicine on Lapicque's theories." I protested and replied, "Sidney, you know that I practically cannot speak English." To which he replied, "Well, you still have two weeks before the meeting and if you can't learn English in these two weeks then you had better go back to Russia."

So I wrote my paper in French and asked a friend, Teddy Eirenrich, to translate it into proper English and read it to me so that I could transcribe it phonetically into Russian. This is how I read my paper at the Academy of Medicine. Needless to say no one asked me any questions, but Sidney was jubilant. His response was, "You see, you could learn English in a couple of weeks."

As mentioned, I was assigned to Dr. Hans Strauss, the Director of the laboratory. Dr. Strauss was a very knowledgeable and kind man. In addition to being well trained, he was a passionate musician and often participated in playing classical music. In his laboratory I was more than welcome. Hans told everyone that in looking over his shoulder during EEG readings I played the part of his conscience so that he became more careful in rendering diagnoses.

While working with Hans we carried out two collaborative studies. The first was related to the comparison of hyperventilation effects in non-epileptic children, and in adolescents as a function of age, and with children presenting with petit mal epilepsy. It was remarkable that both of the curves, as a function of age, were practically identical. This finding suggested that in addition to the components of the mechanism of seizures, the basic chemistry of cerebral vasomotor activity plays its own role.

The second collaborative effort was somewhat more exciting. I found in the records of Dr. Strauss proof that there was often bilateral differences in amplitude of output from the two hemispheres, with the lower amplitude from the left parietal-occipital derivation. This was particularly true for neurotic patients. This suggested that the left hemisphere was more responsive to mental stimulation than the right. This observation was confirmed many times in my subsequent studies. Also during drowsiness the left parietal-occipital region retained its awake pattern longer than the right. Furthermore, in patients with slow wave activity resulting from shock therapy, with bilateral electrodes, the left hemisphere in its posterior region resisted by its activity, the invasion of the slow waves.

Of course this attempt to differentiate the function of the two cerebral hemispheres by the EEG, predated Sperry's work concerning the differential right and left brain function by the use of the split brain preparation. In passing, it might be of interest to note that Beritoff predated Sperry's work in quite a different, and I should say more significant way, than the consideration of only the verbal capabilities of the two hemispheres. Beritoff conditioned pigeons while one of the eyes was blindfolded. This conditioning did not effect the other hemisphere when he shifted the side of the blindfold. Regrettably this remarkable contribution fell into oblivion before the observations of Sperry. Along these same lines new behavioral studies have been made possible by software which was developed to translate wave frequencies into colors on the cathode ray oscilloscope. With this technique it is alleged that differences are observed in the color patterns (wave frequency) in certain psychiatric conditions.

Everyone was kind to me in the laboratory and made me feel at home. I learned some English, although some of its peculiarities escaped my understanding. For example when our technicians would praise me for something and they would say "Oh Boy" I felt compelled to say "Oh Girl." So they laughed at me.

One day I found an EEG record on my table from the preceding evening. I thought I had to read it, but I could not find a history of the patient. The brain waves were uniformly slower than usual, so I wrote: Diffuse Abnormal EEG. Later I learned that the record was from a rabbit and was left intentionally on my desk as a joke.

As I continued enjoying working with Strauss, I found out that Dr. Charles Burlingame, Director of the Neuropsychiatric Institute at Hartford, Connecticut had approached the Rockefeller Foundation to help him establish an EEG laboratory at his institute. Jasper, being of course totally engaged in EEG work in connection with Dr. Penfield during surgery on epileptics could not devote more time to their project than a consulting visit once a month as well as being an initial advisor to buying the necessary equipment. He asked me to take the job. After many meetings with many dramatic incidents mostly related to my poor knowledge of English, the deal was concluded and the Rockefeller Foundation agreed to support my salary for the first couple of years. I was duly notified and with mixed feelings left Mount Sinai with the blessings of my supervisors. It was reported to me that a few days after my arrival, Dr. Burlingame was in Montreal and bitterly complained to Jasper about my English. However when he came back to Hartford one week later, I gave a lecture with enough English to be retained on the staff. Dr. Jasper was most helpful in all this.

When the time came to leave for Hartford I hated to leave New York, which I liked more and more, to bury myself in a relatively small New England town. Yet I welcomed the opportunity to investigate the brain waves of mental patients. My experience with epileptic patients suggested that we might find special EEG patterns characterizing different mental diseases. The latter could be grossly divided into organic cases, e.g. drug induced changes or arteriosclerosis and syphilis, from those without any known organic basis.

U.S.A.

I was advised before assuming my new responsibilities to spend a week or two in a large mental hospital in Worcester, Massachusetts to acquaint myself with the workings of American psychiatry. I was given a room there and was allowed to attend all teaching sessions. I must confess that my sleep was far from peaceful as the dormitory was located near a pavilion of agitated patients, and their incessant shouting distressed me no end. But I was rewarded by listening to the lectures of eminent psychiatrists and attending interesting seminars. In one of them I learned of an extraordinary experiment with hypnotism. It was carried out by a research psychiatrist using subjects recruited from the incoming class of residents. All of them were told under hypnosis the following story:

You were invited to a hospital party where doctors and nurses were enjoying a social get-together. You found yourself sitting on a sofa near a beautiful nurse who was smartly dressed. It was obviously an expensive new dress which probably was bought expressly for this party, probably spending a sizable portion of her bi-weekly salary. You were chatting, smoking and laughing with her. Suddenly you became aware that your cigarette was touching her dress and that you had made a sizable hole in the material. You apologize profusely but the girl was upset, started to cry, cursed you and left the party in distress. Then the hypnotist told them not to talk about the story and de-hypnotized them.

The next day most of the new residents quit smoking. When asked why they could not give a plausible answer. This remarkable experience suggested strongly that we are indeed greatly influenced by the unconscious. Later in my life I experimented a great deal with animal hypnotism which I will discuss later.

At the Hartford Neuropsychiatric Institute, I inaugurated research on mental patients under the name of Functional Electroencephalography. As mentioned before, the Institute was headed by Dr. Burlingame, a very able psychiatrist/administrator and a colorful figure in American psychiatry, and who was the former director of the American Hospital in Paris. He was a superb administrator and succeeded in transforming the ailing Hartford Retreat into a flourishing academically recognized hospital. I never regretted having spent seven years at his institution. He was very generous toward me and appreciative of my needs.

In Hartford, besides me, was an EEGer who was related to Pauline Davis, the wife of Hallowell Davis. He was Charles (Chick) Stevenson who was a competent EEG man. He learned that I had arrived in Hartford and subsequently paid me a visit. At that time there was an EEG society in London, the meetings of which were contributing to the development of EEG in Great Britain. There was also a club of EEGers in Boston. I expressed disappointment with this state of affairs and asked Chick if he would help me with the organization of an EEG society. He readily agreed and I asked Dr. Burlingame whether he would object if we called a special meeting of the EEGers in the Institute of Living. Dr. Burlingame welcomed the idea. Chick and I compiled a list of people who were active in the field and sent letters and made numerous telephone calls inviting them to the Institute to discuss the formation of a local society that we decided to call the Eastern Association of Electroencephalographers. My definition of Eastern at that time was everything east of us and since the world is round this would be all inclusive.

And so the first meeting was held in 1946; thus shortly after the end of WWII I had the pleasure of meeting most of the active EEGers in America. There was Jasper whom of course I knew. There was Gibbs who was very kind in welcoming me to this country. There was Mollie Brazier from England, whose papers were always meticulously prepared and delivered. There was Bob Schwab, a very extroverted neurologist from Boston who was always trying to convince you of something. There was Bobby Cohn from Bethesda, Maryland who was still on active duty as the electrocephalographer at the U.S. Naval Hospital in Bethesda. With this nucleus the EAEEG was established.

Dr. Burlingame belonged to the group of psychiatrists who were organically oriented. They firmly believed that psychoses such as schizophrenia, manic depressive, etc., had some structural (microscopic or chemical) dysfunction that was responsible for the disordered behavior. I most fervently hoped that some EEG findings would somehow buttress their views. The fact that epileptic patients had striking EEG changes despite the failure to find discernible lesions in the brain at autopsy certainly was a boon to these hopes.

The other camp claimed that the main functional psychoses reflected truly non-structural central nervous system disturbance. They hypothesized that these aberrations were due to emotional conflicts in early life or even infancy. These psychoses would be cured when one miraculous day one could communicate with the unconscious of the patient and abolish abnormal thoughts.

My contemporary electroencephalographers working in the mental hospitals such as Morton Rubin and Pauline Davis and others, did reveal some non-specific, very modest deviations in some psychotic patients, but were uniformly unsuccessful in finding specific, or striking abnormalities. As a novice in this field my tactics were guided by the attitude of "wait and see" while collecting a great number of records from patients afflicted with the different psychoses. Some unusual patterns were recognized, but none that were uniformly correlatable.

One cause of difficulty had to be ruled out: the effect of age on EEG. Indeed, whatever pattern one chooses, during the alert resting state and during drowsiness or sleep or other stimulation during testing, there were obvious age differences. The problem was that the incidence of major functional psychoses were greatly influenced by age. Indeed the psychopathic personalities dominated the twenties; catatonic and simple schizophrenia at still earlier ages; in the thirties paranoid schizophrenic patients were prominent. In the forties and fifties depression and manic psychosis become prominent. At the present time there is discussion that depression has no age of prominence except for "involutional melancholia" which seems to be obviously age dependent.

Of course EEG abnormalities were the rule in organic psychoses mostly in the arteriosclerotic and toxic metabolic states. So a different attitude prevails among EEGers working in mental hospitals; namely an obligation to detect organic factors in patients. For example it was essential to differentiate depression from organic psychosis even though some of the depressed patients appeared confused. Our contribution together with psychometricians, was to detect and quantify organic factors and differentiate organic from functional disturbances. Yet our research results were disappointing. Then a thought came into my mind: mental patients *react* to environmental change in an aberrant way. Maybe EEGs that are normal "at rest" may show peculiarities when the patient is subjected to various environmental situations. As noted earlier, I proposed the term of "functional electroencephalography" for the technique of stimulating patients with light flashes, sounds, verbal inquiries, or even plain sleep. I also included a series of brief sleep, as I noted that some patients, particularly depressed and manic, had intervals of sleep of only 2 sec in durations. (I called this "microsleep." More about this term later.)

I was rewarded by the fact that patients with anxiety or depression, reacted under stimulation with more prolonged disturbance of alpha frequency activity than patients devoid of this symptomatology. In other words the EEG objectively confirmed what could be seen by clinical observation It did not reveal new processes which could generate gains in the understanding of the "disease" in the same sense as EEG in epilepsy. I was quite desperate. I could not have an objective basis for brain dysfunction in psychotic patients. On this basis I had to conclude that the chemical processes responsible for the psychoses did not have an electrical expression in the scalp surface EEG. Was I wasting my time? Probably not, because we could help by objective tests, to differentiate organic from functional related diseases. Later new vistas opened as the result of evoked potential studies; these will be discussed in some detail later.

There were two areas where I could be immediately useful as a clinical neurophysiologist. The first involved therapy; the second involved animal experimentation.

Shortly before I came to America electric shock therapy for psychotic patients was introduced. It consisted of eliciting a generalized convulsion analogous to those of "idiopathic" epilepsy. The convulsions were generally followed by variable degrees of memory dysfunction.

This form of inducing a convulsion resulted from the experience gathered in the earlier use of chemical activators of seizures. In the chemically induced seizures following a series of treatments it was observed that depressed patients were able to regain essentially normal behavior. Mechanism of the improvement could not be established, but it was obvious that an electrical storm was triggered. The storm was certainly associated with all sorts of accompanying and underlying chemical functions, but also the elimination of unpleasant memories might play a significant role in recovery. In the elucidation of the mechanism of recovery, however, one could electrically determine whether it was the convulsion, or the loss of memory that was most important. Initially the electric current used for shock therapy was the same as that used in domestic circuitry, namely 60 cycles per sec AC. All I had learned in Lapicque's and Bourguignon's laboratories as to the time factors in stimulation of excitable tissue was against the use of such a preposterous excess of energy in electric shock therapy; I revolted against this type of physiology. I decided to test in guinea pigs the optimal characteristics for the induction of generalized seizures. I went to the Chief Physiologist and explained to him that the amount of electrical energy pumped into the brain of innocent mental patients was too great, and I tried to get his support for my animal experiments. He looked at me with his big eyes and round face and said, "Doctor, it is quite understandable that coming from France, a poor country at war, you feel strongly about saving electrical energy. But this country is, thank God, at peace. We do not worry about the cost of electrical energy for our patients. Why if the referring physicians hear about this, the whole shop will close down and they will stop sending patients. Please do not worry about the cost."

I explained to him that I was not really worrying about dollars and cents, or as a matter of fact, that the new shock machine that I envisioned would be much more expensive than the present one. I sounded convincing in the advisability of my research and with his blessing I started my experimental work.

As anticipated the usual parameters of electric shock therapy were far from being the most desirable to elicit a generalized convulsion. The technique used at that time consisted in delivering a continuous succession of pulses of electrical current at 16 thousandth duration each during a period of only a quarter of a sec. The jolt produced by this current was sufficient to fracture a high percentage of patients' vertebrae and set up a self-sustained electrical storm in the brain which recruits an increasing number of neurons and finally brings the brain to the brink of an explosion resulting in a generalized convulsion. It could be shown that the same type of convulsion could be obtained by using much briefer stimuli or by prolonging the total duration of stimulation. By decreasing pulse duration one relies on the natural property of cerebral tissue to act within a thousandth of a sec. By prolonging the total duration of the stimulation one pushes the brain to achieve a high degree of excitability without a great explosion. Thus the guinea pig gave an unequivocal answer to my question, but doubt remained as to its applicability to man. Despite the fact that the current which I advocated for convulsive therapy was milder than that of the present operating instrument, our Psychiatrist-in-Chief did not want to subject his patients to the test but he had a friend who was the Director of a State Hospital, in the other corner of the country, who allowed me to test the new instrument in his hospital. So I took my machine and my technician across the country for testing. It was a successful operation. I induced Dr. Franklin Offner to build me a brief stimulus electroshock machine which changed the routine electric shock treatment technique by decreasing pulse duration of the stimulus (increase of voltage) and increasing total duration of stimulation (decrease of voltage). This technique was officially approved by the 1st International EEG meeting held in London in 1948.

As is well-known, recording of brain waves is not only a subject of research but is an important diagnostic tool. When I first started to practice EEG, the diagnostic importance of EEG was much more appreciated than now.

Abnormal brain waves were mostly characterized by their abnormal frequency. Thus instead of being in the alpha wave range they became slow, exhibiting either theta frequency (4–7 c/sec) or delta rhythm (below 4 c/sec). As mentioned before, 3 c/sec activity is a very characteristic abnormal rhythm when it is explicit, although it is hidden and of low amplitude in a normal record, being the basic rhythm of the brain (Pi rhythm). Of course the slow rhythms may be normal in young children and infants. The abnormality may be generalized unilateral, bilateral, or focal when it predominates in a small sector of the brain (also unilateral and bilateral).

Thus for example encephalitis, toxic encephalopathy or organic senility are usually associated with diffuse slow activity. In these cases the abnormality is more specific, although in either case it is specific. The abnormality may be focal and then again two possibilities are present: a non-specific focal abnormality, such as the presence of slow waves over a limited portion of the scalp; or specific abnormality, also called paroxysmal abnormality characterized by either high amplitude waves, or spikes, or spike and waves. The more specific focal abnormality may correspond to the presence of brain tumors.

When I started my diagnostic work the diagnosis of brain tumors by EEG was an important dreaded possibility. Unfortunately when I became proficient in the diagnosis of tumors, it was one of the saddest experiences of my professional life to make this diagnosis as the prognosis of brain tumors was quite grim. At the present time there are many other more precise techniques of diagnosing tumors and the electroencephalogram is less often depended upon for the diagnosis. On the other hand, the "paroxysmal" focal or more focal abnormalities are most often related to epilepsy and in this area the EEG remains a most valuable diagnostic tool.

I remember one night I was sitting in the second row at the Metropolitan Opera watching a ballet when a man sitting just in front of me suddenly emitted a characteristic cry and started to shake. People sitting nearby as well as those on the stage understandably became concerned. I whispered firmly not to be alarmed, placed my hands on the shoulder of the shaking gentleman and succeeded in pacifying everyone so that the show could go on. I knew it was a typical grand mal seizure and that in a minute or two it would stop. Indeed, grand mal attacks start with a cry, loss of consciousness, generalized contractions, followed by repeated shaking of arms and legs and is often followed by a period of relaxation and often sleep. It is usually preceded by an "aura" the significance of which the patient learns by experience; it may be a particular perception or a movement or even a thought. Present in this fashion the epileptic attack is one of a great variety of typical spells. Let me review succinctly the most frequently observed ones:

a. A "petit mal" is a sudden onset of unconsciousness which may occur without any aura. It is usually observed in children and adolescents. Suddenly, while the child is having a spell he/she becomes aware of having missed some thoughts or movements. An observer can perceive a sudden immobility of the face, lasting a relatively brief time until consciousness returns. Attacks of petit mal may be rare or frequent and in a number of cases they precede the onset of a grand mal seizure.

b. A psychomotor attack also occurs without being associated with generalized convulsions. The patient may suddenly start moving his/her mouth in a peculiar way and then say or do something quite peculiar but always the same during repeated attacks. They would not remember what they were doing or saying. This is a general characteristic of epileptic attacks. Thus the patient does not remember the convulsions of the grand mal but is aware of its occurrence either because he finds that his recollections are missing or that he wakes up from an unexpected sleep.

c. Finally, an epileptic attack may be "focal," e.g., one hand shaking or one hemiface repeatedly contracting.

These manifestations of epilepsy are so vivid that they were known in antiquity and were interpreted in most mysterious ways. With the knowledge of medicine the story was complicated by the fact that patients having either a general grand mal or a petit mal seizure have no demonstrable lesions in their brain. However, those who have psychomotor or focal seizures may have had a head trauma or manifested a definite brain lesion. It was H. Jackson, a British neurologist and Bravas in France who identified the relationship between specific brain lesions and characteristic individual epileptic manifestations of epilepsy.

Now imagine against this background of certain forms of the mysterious disease such as generalized grand mal and petit mal, having no objective correlate in brain tissues examined during autopsy of patients who suddenly died from other causes, a sudden light was projected by electroencephalography which facilitated the diagnosis, the prognosis, the treatment, and most particularly the objective brain disturbance correlated with this illness. Indeed, as soon as EEG could be recorded from the scalp, patients with epilepsy revealed unmistakable and specific electrical patterns.

At first it was Gibbs working with his wife Erna who described a remarkable and constantly recurring pattern in patients with a petit mal epilepsy. It consisted of bursts of high amplitude 3 c/sec wave and spike formations. This remarkable pattern is the signature of a diagnosis of petit mal. Conversely, its absence when EEG is recorded during the spell rules out this diagnosis. We again remind the reader that 3 c/sec or Pi rhythm is the one that I consider as a basic rhythm of normal EEG provided that it is of too low amplitude to be recorded in the actual tracing. However, in patients with petit mal epilepsy its amplitude is so great that the pattern is easily identified. It was my privilege to have been the first to record a DC potential (non-oscillating potential) underlying the spikes and waves. Several years later Cohn made the same observation, being unaware of my earlier "discovery." Then many electroencephalographers observed spikes over the DC potential in the records of epileptics, particularly those with or without head traumas, although in some of these patients, focal bursts of high amplitude slow waves may characterize the presence of an epileptic lesion.

Finally, in patients with grand mal characteristics electrical patterns were also recorded. These are characterized by a sudden increase of the amplitude of the brain waves as well as a change of their frequency. During the initial "tonic" phase the high amplitude, high frequency pattern is continuous. During the "clonic" phase, the burst of high amplitude waves and spikes are interrupted by a flat line corresponding to the intervals between the convulsive movements. To sum up, the clinical seizure corresponds quite closely to an electrical disturbance, interrupted during the clonic phase.

For the first time in the history of medicine, brain electrical recordings objectivized an illness which until then was lacking any objective correlations in brain studies. These exciting findings of course stimulated the imagination of psychiatrists who expected that EEG would show specific electrical manifestations of functional mental disorders. Unfortunately, nothing of the sort happened—and yet we tried.

The possibility of recording objective electrical manifestations of epilepsy was helpful in the case of febrile convulsions of children. It is well known that some children convulse while having a high fever. Electroencephalography was helpful in such cases. I found in some of these cases definite unilateral abnormalities. In other cases there was no trace of epileptic pattern when fever was down. Thus EEGs were very helpful to parents worrying whether or not their child would have convulsions in the future.

Two schools of thought were particularly involved in the study of epilepsy in America. Doctor Penfield, a neurosurgeon in Montreal working intimately with Herbert Jasper, an electroencephalographer, held the view that epileptics have two different etiologies of their illness. Those with petit or grand mal with symmetrical manifestations were found to have a disturbance in the deep mid-region of the brain that caused this disturbance. The others must have some focal brain lesion either unilateral or bilateral that may be susceptible to cure by surgery.

On the other hand, those who initially worked with Gibbs, without denying this possibility, were much less surgically inclined. I had the opportunity of learning from both, so that when I opened a seizure clinic at St. Francis Hospital in Hartford, Connecticut, I learned a lot about this illness and these patients. One must be aware of the fact that epilepsy may not interfere with a patient's intelligence as many great men were known epileptics, for example Dostoevsky and Caesar.

Every practicing physician has observed emotional factors contributing to the onset of epileptic seizures. Two examples follow: In one case an epileptic female had been seizure-free for three years. She brought to my laboratory her 14-year-old daughter, who after eating green apples had a first seizure. EEG records were obtained from the mother and child who had almost identical high amplitude spike and wave formations which were consistent with a seizure disorder. When I reported these findings to the mother she promptly had a major seizure, which was precipitated by violent emotion. In the second case the seizure occurred during insulin shock treatment. Her physician, who was making rounds, sat on the edge of her bed, took her hand and gave her a kind professional smile. A grand mal seizure followed, apparently induced by a strong emotional stimulus.

Obviously major seizures may be produced by any modality of sensory input. Numerous investigators, following the initial observation of Grey Walter, have produced epileptic seizures associated with visual stimulation. Therefore the sensory stimulation and psychological excitation may be interchangeable either in the same individual, or in individuals who have a low threshold to activation of the synchronous depolarization process. These phenomena bring up an important point to be answered. Are we justified in distinguishing the laws and facts of a psychodynamic process from the laws and facts of a neurophysiological process? Are we able to link these two types of detonating agents? My notion is that in each instance the elementary mechanics are very similar. In psychodynamics, conflict is a basic problem. Sherrington proposed the notion of a fundamental conflict between the innervation of antagonists (flexorextensor) in muscle action. Psychoanalysts indicate that the most common solution of emotional conflict is repression. Sherrington found that the essential principle of spinal cord coordination is reciprocal innervation, the principle which resides in reciprocal inhibition. According to this principle, each time a flexor reflex is activated the extensor reflex is actively inhibited. It seems that the phenomenon of inhibition plays a fundamental role in all classical neurophysiology, including the physiology of Pavlov.

We can envisage that the solution of psychological conflicts may be accomplished by various mechanisms, such as sublimation, rationalization or substitution. The substitution mechanism was described by Uchtomski under the name of "Dominant" and may be observed in the spinal frog. According to these experiments a slight stimulation of the sensory nerves, which usually elicit flexor reflexes, sensitize the flexor centers to the point that excitation of any other nerve continues to produce flexion instead of extension.

It is said that psychological activities are essentially motivated consciously or unconsciously. However, if the conscious motivation is absent, of course on a neurodynamic level one should find in a sort of intermediate stage, the elements necessary for the development of the psychological motivation in simple neurophysiological facts. This is the point which should be stressed. We usually say that reflex functioning is represented by a circuit which is opened at the periphery. We discuss an arc reflex. A stimulation produces a central perturbation which in turn is followed by centrifugal action of excitation which elicits the motor act. One tends to forget that the motor act in turn generates centrifugal messages; one may call this phenomenon of closing a reflex circuit at the periphery, the peripheral closing. Such peripheral closing is present in the simplest spinal reflex because even a muscle detached from the joint is the site of origin of stimuli derived from intramuscular sensory fibers during reflex contraction. At higher levels of the neuraxis this phenomenon achieves greater importance. This is the dominant factor in socalled chain reflexes in that each segment of the chain elicits a stimulus, which in turn drives the next segment so that a complex action may ensue. At the level of conditional reflexes peripheral closure (or confirmation) is essential to the persistence of conditioning, which otherwise would be inhibited. In some later work I showed the importance of peripheral closure (or peripheral confirmation) in the execution of voluntary movements of the simplest kind, where the subject produced rhythmic extension of the fingers as rapidly as possible. These movements were irregular and relatively slow. In carrying out such movements when the dorsal aspect of the finger is allowed to touch a sheet of paper immediately the rate of movement becomes much more rapid, and the action becomes more regular. Under the initial conditions the movements are handicapped by the failure to recognize the fact that the movements were properly accomplished by the difficulty in achieving peripheral confirmation. By placing the paper, the signal of accomplishment of the movement is rapidly achieved and the operation is considerably facilitated.

It is important to have a clear vision of the importance of peripheral perceptions in neurodynamics because this phenomenon is the immediate precursor of those that underlie psychological maturation. Imagine that I wish to touch a table in front of me with the tip of my finger. When I do this movement I am not aware of the contraction of the muscles of the shoulder, arm, forearm or hand which all participate in the movement. My consciousness is concentrated on the tip of my finger and the anticipated sensory perception. It is as if the movement was motivated by the anticipation of sensory stimulation. If one attempts to imagine how this anticipation might be realized in the brain one does not have any choice but to admit that the *goal* of the movement is represented (modeled) in our cortex. The following clinical observations on the phantoms of amputees confirm this viewpoint. It is well known that many amputees may not only have illusions of moving their amputated extremity, but they may have the faculty of perceiving the results of such imaginary movement. They may for instance have an illusion of touching an object by trying to move their absent leg; they therefore stimulate directly the neuronic circuit of the brain which creates the impression of a "real" perception. The following hypothesis may be formulated: Each time that one pursues a certain goal to execute movement, we activate intracerebral neuronic circuits which are a kind of phantom of the movement. This phantom, the principle part of which, if not the totality, is presented as an image of the goal to be achieved. I am aware that this formulation has some problems in that topectomy does not, in general, relieve painful phantoms, but this latter may be due to the failure to know the precise anatomical organization of the phantom "area."

In normal function these phantoms are not perceived as being inhibited by "real" perceptions. We now find ourselves at the important cross point between neurodynamic and psychodynamic mechanisms, particularly perceptual significance. We are now meeting the problem of localization. One often says that neurodynamic mechanisms are intimately related to precise anatomical sites, while this is not the case for psychological phenomena. This is not exactly correct. The principle of reciprocal inhibition of the peripheral closure of substitution of conditioning is observed at many levels of the central nervous system. One of the most fascinating conquests of modern neurology was to find that the condition of consciousness (the state of propositional awareness) depended on the integrity of the brainstem, particularly in the region around the aqueduct. One also found characteristic personality changes with destruction of thalamic-frontal pathways. Dr. Penfield found that the electrical excitation of the temporal lobe of certain patients might produce recall of very complex "dream-like" phenomena. Actually the dichotomy of neurodynamic and psychodynamic is fundamentally a semantic exercise. Whether we are discussing the neurodynamic or psychodynamic phenomena the analogous laws may be described provided obviously, that one considers the different unities to be governed by these laws. The unities are quite complex in psychodynamics, whereas in basic neurodynamics we deal with muscles, the "simple" motor complexes or sensory perceptions. In psychodynamics the presence of complex social activities and their affect producing properties via the delicate nuances of verbal communication become manifest. The way in which these complex behavioral "units" are represented in the brain is very difficult to imagine, but not too much different that to understand how one can perceive a table through the tip of a pencil. In each case we cannot have a precise concept to be formulated as to the formation, or conservation, of these complexes which constitute the "units" of our behavior. However if we accept that these complexes are somehow represented in the brain, then it is evident that the dynamic reciprocal relationships between them are analogous.

Let us see how such an integrative concept may effect the comprehension of psychopathological phenomena. One becomes acutely aware how the clinical diagnostic "entities" are fuzzy, artificial, nosologically incomplete and unrelated to the remarkable variations of human behavior. Thus if one decides to consider, instead of diagnostic labels, the isolated symptomatology, one appreciates the aid which may be derived from the knowledge of neurodynamic principles. Let us consider mental disease in a sense as wide as possible, in terms of neurodynamic processes such as excitation, inhibition, substitution, peripheral closing, communication, memory, etc. Let us ask ourselves, for example, what would happen to the human mind (the mechanism whereby sensing, comparing, categorizing, symbol operations and integrative processes take place) if any of these basic neurological mechanisms were disturbed. Imagine, for example, what kind of clinical situation would result from hyperexcitation of the cognitive elements, or of the language processing operations. The world and its symbols might be more REAL than reality itself and so-called delusions and hallucinations would invade and alter, or destroy, logical brain operations.

Our analysis of neurodynamic principles led us to the existence of phantoms (models) of real behavior in our mind (brain operations), and the appreciation of our reactive universe. It would be sufficient for any major (or even minor) elementary dysfunction to alter organismal behavioral patterns and to generate symptoms of "mental" illness.

CHAPTER 4

Psychiatry

*"But enchanted and bewitched
How long would this state last
Can the unreachable be reached
Isn't the present already the past?"*

As NOTED EARLIER, I assumed my responsibilities at the Neuropsychiatric Institute of Hartford, Connecticut, otherwise called the Institute of Living. The residents called it the Institute of Loving. My immediate goal was to attempt to find correlations between different types of mental disorders and EEG. I attended many diagnostic sessions at which the patients were discussed according to the jargon of mental hospitals. Often, it almost seemed as if for each individual patient there were as many diagnoses as there were individual psychiatrists. However, the guidelines were there.

At first there was a differentiation between psychotic and psychoneurotic patients. The primary difference was that in psychotic patients, confusion, hallucinations, delusions and false beliefs were paramount. Among psychotic patients, the chief differential was between schizophrenic and manic-depressive, involutional melancholia, arteriosclerosis and finally toxic psychosis. Here the main subdivision was between indications of life-long psychotic personalities versus isolated definite precipitating factors following which the patient might manifest confusion, hallucinations and delusions. The first group were called indigenous psychotics, the second were called reactive psychotics. The prognosis for reactive psychotics had a more promising outlook for recovery than the indigenous patients. Among the psychoneurotic patients were the anxiety reactions, depressive reactions, or obsessive compulsive and agitated depressions. These latter patients were overactive, somewhat agitated and usually had difficulty sleeping; but had no signs of confusion, delusions or hallucinations.

Active depressions were those who had given up the struggle for life, some of whom were acutely suicidal. Obsessive compulsives were typified by patients who washed their hands nearly every hour for fear of contamination from touching objects around them and those who would repeatedly return to their abodes to check the doors, light, etc. But of course the combination of these symptoms resulted in a diagnosis of mixed psychoneurosis. Involutional psychosis involved patients over the age of fifty. These patients were extremely anxious, often confused and sometimes paranoid. In such cases they were called paranoid involutional melancholics. Schizophrenics were patients who had lost contact with reality. Sometimes they were religious fanatics, believing they were carrying God's messages, sometimes agitated with stuporous posturing in which case they were called mute catatonic schizophrenics; sometimes convinced that they were receiving messages with hostile intent, e.g., their relatives would like to kill them and their friends were betraying them. When they had these acute delusions they were called paranoid schizophrenics. The manic-depressives, when in the manic phase, continually talked, were continually agitated, their verbal fireworks seemingly inexhaustible with gesticulating, shouting, confusion. When in a depressed state they were often self-destructive. Arteriosclerotic patients were early observed as clinic outpatients. They were frequently disoriented in time and space, often mumbling and telling unbelievable tales. Later they would become incontinent. At times if their verbal output was analyzed they betrayed personal identification with important life events. Many of these patients are now identified with Alzheimer's disease and associated processes. We also studied toxic psychoses of chronic alcoholism and users of illicit drugs.

And so it was my specific task to determine whether these classes of patients showed outstanding or readily defined patterns of electric activity. At that time it was the dream of electroencephalographers to find EEG patterns that paralleled those of the epilepsies. This dream was destroyed as the number of patients increased but no characteristic patterns with any of the psychiatric nosological entities were found. A simple eye-ball way of determining generalized abnormalities such as decreased basic frequencies below eight per second and the presence of generalized bursts of theta or delta activity of the brain were of course observed, but no specific patterns were recognizable. Moderate electric abnormalities

Psychiatry

were characterized by eruption of theta rhythms primarily in the anterior temporal region, then the minimal deviations such as contour and amplitude variations from normal; and then of course normal records. Theta activity in the occipital regions was also considered as a major abnormality in our classification.

It appeared that there was a greater number of abnormal patterns in mental patients than in control subjects. The percent of the various abnormalities ranged around 10% at most and certainly did not relate well with the severity of the psychoses. We also had to determine the most effective electrode ensembles for standard recordings. Since the early work of Berger and Adrian it was known that alpha frequency activity predominated in the posterior quadrants, sometimes with decreased amplitude on the left. The activity in the frontal regions was much less well organized, with alpha waves being mixed with a great deal of beta activity and even some ill-defined theta waves. Temporal regions showed an intermediate pattern.

We soon acquired a six-channel apparatus which permitted us to record simultaneously from six different regions. As far as 4 channels were concerned, there was no hesitation in separating posterior occipital and anterior frontal outputs. It is appropriate to state here that there was an overt controversy among electroencephalographers, particularly between the Gibbs' and Jasper as to the use of so-called monopolar and bipolar derivations. Gibbs favored monopolar and Jasper bipolar. It is obvious that there is no absolute monopolar derivation as one is always measuring potential differences between two "points." In the monopolar derivations it is assumed that one point is inactive. The arguments were really quite acrimonious and generated little light. Douglas Goldman argued that to get around these difficulties one should use an "averaged point" by connecting all the electrodes together as the reference. This method became popular for a time. When I had 16 channels at my disposal I demonstrated that "monopolar" leads are not those that express focal abnormality, but they do express epileptic spikes. On the other hand, bipolar electrodes are more effective in polarization of brain waves, a result that was not exactly expected from the many decades long controversy.

But the big question remained, was there any difference between different psychiatric groups? If not, was there the presence of specific patterns? The final answer was yes; however differences were such that age effect could not be eliminated.

Indeed, electroencephalograms mature and involute. At the time of my taking charge of the EEG laboratory at the Neuropsychiatric Institute, many papers were suggesting that the mature pattern in the posterior leads were not achieved until the age of eight. However, we convinced ourselves that the maturation continued until the age of twenty. Our younger patients, regardless of diagnosis, had a great deal of theta activity in the anterior and middle regions of the brain as well as some peculiar occurrence of isolated slow waves in the occipital regions. In middle aged patients there were often bursts of theta activity in the posterior regions and a great deal of fast activity in the anterior regions. Of course, in older patients the basic frequency was often below 8 per second. Involutional melancholics in particular were characterized by changes found in old age. It was therefore our uncomfortable feeling that if the different diagnostic groups were characterized by different ages, then we could not be sure whether the patterns which were found in the different groups were due to age or to mental disease. Very quickly we convinced ourselves that our premonition was correct. Schizophrenics of the catatonic or simple type were mostly below 25 years of age. Paranoid schizophrenics were usually in their thirties; manic depressive, depressed or manic, were in their forties and involutional melancholics were in their 50's and 60's. Arteriosclerotic patients and senile patients were over sixty. It was difficult under such conditions to determine what was the result of mental symptoms and what was best correlated with age.

Two phenomenon stood out; how the patient's EEG responded to various stimuli and how compelling was the differential tendency to sleep during day time examinations. There were also differences in the sleep patterns of the different groups. It became very clear to us that it was foolish to try to determine a differential pattern from the resting EEG alone. However, the EEG provided us with an index of the electric activity of the brain to various different stimuli. I named this "Functional Electroencephalography." I described it in the following way. Although two identical records may be obtained under resting conditions in two different patients, the EEGs may be quite different during stimulation and/or during sleep. In some patients there was no change of the EEG under the influence of stimuli, such patients were mostly arteriosclerotic, or toxic psychoses. In other patients the depression of alpha frequency activity not only followed the first flashes of light but lasted during the entire experiment. Our very last patient did not accommodate to photic stimuli. Actually, there was a differential response to light in a number of patients; some showed transient responses, others showed responses primarily over the left occipital region. Similar responses were found with other modalities of stimulation, for example, peripheral nerve stimulation or having the patient carry out arithmetic operations and other mental tasks.

Variable sleep patterns were particularly impressive. I can remember this was one of the most fascinating experiences of my early EEG studies. From the very onset of the slightest cloud of drowsiness, the EEG showed the characteristic depression of alpha activity in the occipital regions and the occurrence of theta waves in the anterior and middle regions of the scalp and then suddenly, sharp waves of high amplitude which Gibbs called "parietal humps." Later V waves appeared which predominated in the vertex regions. The next stage was characterized by the appearance of 14 per second activity (spindles) in a bilateral distribution. In some patients deviations from the described sequences were observed. The deviations were observed in the involutional age group and the deviations were mostly expressed as 6 or 7 per second activity in the posterior leads. In younger individuals at times slow waves were observed in these regions. Delta activity in older patients at times occurred as bursts all over the head; but the most interesting activity was the presence of slow waves of short duration of around one or two seconds "microsleep," which were most usually found and most clearly defined in patients with manic psychoses.

As noted above, after several years of research in the EEG laboratory of a large psychiatric hospital, it became obvious to me that the expectations of Berger and many electroencephalographers after him to find some specific abnormal EEG patterns characteristic of schizophrenia or manic depressive illness to be futile. At most, non-specific moderate abnormalities were observed in some patients without obvious diagnostic significance, but even in these cases the EEG changes forced us to investigate further for etiological factors. At the present time some brain mappers suggest variable success in multiple correlations of EEG and psychiatric disorders but the real essence of the disturbances escape such a diagnostic tool. Logically it seems likely that some definite objective evidence of disturbed brain function might occur at the time of relapse to psychotic behavior, since a world of images, or phantoms exist between our percep-tions and resultant actions; particularly if there is an unusual chemical neurotransmitter excitation of the neurons sustaining these phantoms. In fact, these phantoms may appear as real to the patients as reality itself. But it should be kept in mind that reality is the result of a consensus of the integration of signs and symbols input from the surroundings and dysfunction may result from difficulty with cognitive discrimination and therefore gross electric abnormalities would be very unlikely.

There is one objective finding in functional psychotic patients that I investigated for years. This concerns word association processes. Usually the word association test of Jung is used to detect emotional conflict. The examiner enunciates a word, either neutral or emotional and asks the patient to respond to it as immediately as possible, with another word that has associational qualities. If the patient is "blocked" for example by the word "suspicion" a paranoid condition may be considered. If the patient blocks for the word discouragement, depression might be entertained. In fact, we found that those words usually block in perfectly normal individuals.

So we chose a list of Jung's 15 most "traumatic" and 15 most "neutral" words. The average association times for these two groups were calculated and resulted in an interesting discovery. The word association time was significantly longer in both traumic and neutral words in patients with "functional psychoses" than in neurotic patients who showed in turn, slightly longer average times than normals. The slowest word association processors were found in involutional depressed patients. However, the slowing was relatively the same for neutral and emotional words, so that the ratio of both average association times was the same for all individuals, the most psychotic and the most normal. These association processes happened to be twice as rapid for the neutral than for the emotional words for all groups of individuals. Therefore what we measured was the rapidity of mental processing. Yet the rapidity of perceptual processes was normal in all individuals with mental illness. This meant that the intrinsic processes of mentation are selectively affected in these patients. In patients with organic mental dysfunction *both* extrinsic and intrinsic processes are slowed; a similar distinction is seen in classically administered electroshock.

While at Hartford we were interested in studying physiological indices other than the depression of the alpha frequency in psychiatric patients. Together with Dr. Miriam Liberson we showed that individuals differed as to their cardiac responses to loud auditory stimuli. These responses were characterized by a shortening of the intervals between two QRS complexes corresponding to increased heart rate and then immediately afterwards a sort of compensatory inhibitory state, characterized by a lengthening between the QRS complexes. In some patients instead of a single biphasic change there were arrhythmic changes associated with the auditory emotional stimulus. Reviewing the entire cardiogram we found that although there were changes in the heart rate there was no change within the QRS complex itself. We concluded that emotional changes are expressed only in the rate, and not by intrinsic systoles of the heart. This gave us an idea, because from the literature and from our own experience, there was a substantial relationship between the duration of the QRS complex and the heart rate under resting conditions. If the heart rate was relatively high under resting conditions, the QRS complex was relatively short; if the heart rate was slow the QRS complex was longer. These latter findings were apparently intrinsic characteristics of the subjects and not related to their emotional state. Hence in our study whenever we had a fast rate in the resting record, before the auditory stimulus, we determined the index between the rate and the QRS complex duration. If this rate was faster than what we expected from the duration of the QRS complexes, we felt reasonably sure that the patient was emotionally aroused. On the other hand, if the rate was too slow for the QRS complex duration then we inferred that the patient was resistant to the auditory emotional stimulus, or was reacting to the stimuli of the environment by a state of inhibition. This interesting observation was communicated in the press but unfortunately this knowledge was not picked up by cardiologists. The bipolar response itself was an interesting observation as it is an expression of a feedback mechanism which must play a fundamental role in the maintenance of stable heart rhythms.

Together with the EEG and EKG, another physiological process, the psychogalvanic skin response, had its own interest. Using standard EEG equipment and placing the electrodes on the palmer surface of the hand it was very easy to make such recordings. In that the standard EEG apparatus has a time constant characteristic because of effective condenser coupling, the time course of these recordings were not high fidelity; they were actually distorted. Instead of a true very slow developing wave, the AC coupling showed a bipolar oscillation. However, despite the wave form distortion one could determine the latency with high precision and could estimate the amplitude fairly well. In the early work the latency values were considered to be an intrinsic brain phenomenon. The measured latency ranged around 1 to 2 seconds, but to our amazement we did not find in the literature any indication that the major part of this latency was spent in the conduction of the impulse from the brain to the skin of the hand. Investigations of Erlanger and Gasser clearly determined that conduction velocity was a function of the diameter of the nerve fibers. The nerve fibers responsible for the potential changes which are related to the activity of the glands of perspiration were definitely among the slowest in the body, so to travel from the brain to the hand took almost the entire latency time. To demonstrate this phenomenon we placed electrodes on the head, over the deltoid muscle, over the dorsum of the hand and the plantar region of the foot; then a loud noise was sounded. The latencies were easily recorded and measured from the successive loci. Our findings were confirmed by a group of Yugoslavian investigators. This important observation was presented at a meeting of the American EEG Society. We were surprised that Dr. Darrow, who was especially interested in autonomic nervous system functions, criticized the paper, expressing disbelief that the conduction velocity was so low. But as I stated our data had already been confirmed. I still believe that the clear meaning of our observations have not yet been effectively explored.

In the meantime, we determined whether there was a relationship between the measured latency and mental disease. We found no relationship, which seemed entirely natural as the latency was nearly all dependent on the peripheral transit. On the other hand, there was a definite relationship between the amplitude of the psychogalvanic reflex and symptomatology and diagnosis of the patient. In patients with organic arteriosclerosis, or toxic psychosis, the amplitude of the psychogalvanic reflex was generally lower than usual for the age. In patients with schizophrenia or neurosis the amplitude was generally within the normal range. Depressed patients, whether neurotic or psychotic, were uniformly of decreased amplitude. It was of interest to find such a symptom that correlated with a functional mental disorder. Of course, much later when depressions were treated with mood elevators which have a definite sympathetic or trophic function, such a response would be expected, but at that time we had not yet entered the era of intense psychopharmacology.

Patients with anxiety showed a greater activity to stimuli; these responses paralleled those seen in the EEG. However, EEG depression was quite developed in psychoneurotic patients and therefore in these patients there was a dissociation between the cortical activity of the brain and the activity of the autonomic nervous system as manifested in the psychogalvanic reflex. We have studied several hundred patients and the analysis which we carried out at that time related only to the INITIAL stimulus. As a matter of fact our technique consisted of delivering 3 intense auditory stimuli with three minute intervals between the first, second and third. Only recently we completed our analysis of the accommodation of the psychogalvanic reflex to repeated stimuli. We found that in psychoneurotic patients, particularly with anxiety, the first response was highest, the second lower and the third lowest. In cases of depression and toxic psychosis the response to the first stimulus was already decreased, the response to the second slightly lower, but there was no lowering to the third stimulus. This type of patient did not present a decrease of psychogalvanic reactivity nor a disturbed accommodation process. If one finds any significant correlation between mental illness and psychological indices, one should also find a correlation between the psychophysiological index and age. Indeed, there was a progressive decrease of psychogalvanic reflex to auditory stimuli as the age of the individual progressed.

One of the major applications of EEG sonography at the present time, is the analysis of what happens to certain patients during sleep, generally while they snore. They may "swallow" their tongue, block their respiratory pathways or even die. I believe I was the first to record EEG while the patient snored. It was an interesting case, since the patient's account was vehemently contradicted by his nurse. The patient claimed of not having slept at all during the night. The nurse claimed that each time she visited the patient, he was snoring. We recorded the patient's EEG during the night and found a continuous alternation of alphas and spindles. During the time he had spindles, he was snoring.

After the appearance of V waves, true sleep ensues. Davis demonstrated that the V waves were evoked by auditory stimuli. After the introduction of averaging it was recognized that any stimulus would generate V waves. This suggested that in the absence of any stimulation, the response was the result of some internally generated stimulus. These potentials could be confused with epileptic paroxysmal discharges. However, in most cases the paroxysmal bursts of epilepsy had a different distribution. Following the V waves, elegant spindles of 14/sec activity made their appearance. It is interesting that in non-humans 14/sec activity is observed in the awake state. In normal individuals the EEG makes its appearance at about this stage of light sleep.

This slow wave sleep is repeatedly interrupted by the disappearance of slow activity. This disappearance was recognized to be associated with REM and extreme muscle hyptonia. It is during these periods of REM sleep that subjects usually dream. Animal research reveals that the center of REM sleep is located in the vicinity of the nucleus of the third cranial nerve, approximately where Lhermitte and other neurologists located the sleep center. It is remarkable how clinical observations may lead to a precise localization in the absence of experimental confirmation. Yet this appears to be the center of paradoxical sleep, unknown to Lhermitte. The center for slow sleep appears to be much more diffuse and includes some of the thalamic structures.

In a 1965 study at Hines V.A., in Chicago, with my wife Cathryn, we studied the onset of sleep and found it to be marked by slow eye movements always present with the eyes moving majestically from side to side. We found that at the moment of onset of sleep and for a few seconds afterward, characteristic changes occur. Alpha disappears and the subject's mentation changes its character from concrete thoughts to vagueness, with hypnogogic images sometimes reported. The subjects would deny falling asleep, until V waves and spindles appeared over the vertex region. Somewhat later external stimuli elicits K complexes which are much slower than the V waves and predominate in the frontal regions. Then the slow activity invades the frontal lobes, usually in the delta frequency range and with high amplitude. Thus the K-complexes (delta waves plus spindles) appeared to obvious external and hidden internal stimuli.

I have always deplored the omission by most sleep researchers of recognizing the regional differences of sleep activity. I also deplore the tendency of contemporary students of sleep to ignore the obvious rush forward of the total EEG energy during sleep. In the alert state most of the energy is in the occipital region, being depressed more on the left than on the right. I wonder if this decrease in energy during sleep, even outside of dream states may not indicate in some fashion a readiness of the sleeper to respond to visual sensations.

I believe that these spatial characteristics of sleep waves have been underestimated by sleep researchers. While it is generally recognized that slow wave sleep is a deep sleep, I found that it is deeper when the slow delta waves are recorded in the prefrontal areas than when they are recorded in other areas of the brain. Thus I propose a more refined spatial criterion of the depth of sleep. Incidentally, the frequency of delta waves is 1.5 to 2 per second. This happens to be the same frequency as that of peristaltic movements of the intestines. It is also one-half of 3.3 c/sec frequency so important in EEG.

The first lecture that I gave in Hartford was on sleep and dreams. My lecture included a summary of a well-known French neurologist, Lhermitte as well as some of my own studies. In his book, Lhermitte hypothesized that the center of sleep must be located in the brain stem near the nuclei of the oculomotor nerves. Patients with encephalitis, sleep almost continuously and often have paralyzed intrinsic eye muscles. As is well known, Lhermitte's prediction made on clinical grounds was confirmed by direct experimentation on animals.

I gave this lecture to psychiatrists when this domain appeared full of promise in their field. Most psychiatrists have been intrigued by dream research since there is some resemblance of dreams to the hallucinations and delusions of mental patients. Sleep deprivation has also been shown to result in a state of mental instability.

Ever since I first recorded an EEG displaying progressively increasing drowsy and sleep patterns, I became fascinated by the elegance and complexity of these features.

My observations concerning the incidence of drowsy patterns during daytime recording of the EEG in mental patients showed a maximal incidence of these patterns in manic patients, with a lesser incidence in depressed patients and paranoid patients showing minimal incidence. However, this might simply express a compensation for sleep deprivation at night. It is now common knowledge that dreams occur mostly during rapid eye movements (REM). Several attempts were made to determine whether there was a relationship between the percentage of REM and the patient's symptomatology. The unsolvable problem was that percentages of REM and slow wave sleep have a reciprocal relationship and one never knows whether the increase of REM percentage is not due to the decrease of slow sleep as a primary event.

In this first lecture I described a night's sleep as having long periods during which there was no slow activity usually characteristic of sleep. Unfortunately my implicit description of paradoxical sleep was incomplete as I did not associate rapid eye movements with this sleep pattern. As is now well known, REM associated with paradoxical sleep were first observed in children where they were particularly pronounced before they were identified in adults. These observations are consistent with the fact that great discoveries may be missed if they are not associated with careful observations. And so even though I early described paradoxical sleep by its most important characteristic—the absence of slow waves—I did not notice the eye movements and therefore was not sufficiently impressed by my observations to study them separately from slow wave sleep. I did report in the same lecture an example of a patient with the problem of swallowing during sleep.

However, my main description was based on recording of EEGs during daytime examinations. Thus I stressed the fact that brief episodes of sleep invade our wakefulness. In some cases (a second or two) sleep appears, characterized by high amplitude vertex waves or spindles during extremely brief periods of time. I called this "microsleep." This terminology has been universally accepted, although my discovery of it is rarely acknowledged. Many years later, while waiting in a barber shop, I was reading a magazine and to my astonishment I saw the term "microsleep." The article pertained to a famous sports figure who participated in an experiment in which he was kept awake for more than 25 days. This, of course, was similar to manic patients who hardly sleep during the excitement phase of their illness. Later, at one of the EEG meetings a man approached me and said, "Dr. Liberson, I must apologize for using your term 'microsleep'. Frankly," he said, "I thought you were dead." After a brief conversation he divulged that he was the one who had written the article. I said that I hoped in the future he would use my name and he promised to do so.

I also showed, in my first lecture, some unusual patterns of the onset of sleep. For example, the beginning of sleep is characterized by slow activity in the occipital region or it may begin with high amplitude slow delta waves as is sometimes observed in manic patients or patients with arteriosclerosis. I described in detail the onset of sleep, which is preceded by very slow eye movements, later becoming more vague. Thus the EEG may show some theta activity from which emerge V waves, sharp waves of relatively high amplitude predominating in the vertex region. Inasmuch as repeated auditory evoked potentials may induce such waves, I speculated that they constitute responses to some internal stimuli. Unfortunately even today, we still know little about V waves.

Up to this point, if the subject is asked whether he is asleep, the answer will be negative although some drowsiness may be acknowledged. Thus out of the blue episodes of 13 to 14 sec spindles appear also in the vertex region. At that time only a few individuals denied that they fell asleep. At the time of my lecture the significance of spindles were not known. Since then a great deal of research has been done and the spindles are being recognized as of thalamic origin, probably interfering with the perception of the outside world. The significantly slower and more complex slow paroxysmal waves appear more frontally but their significance is also obscure. Finally delta activity invades the occipital cortex being increasingly prefrontal with the increase of depth of sleep.

Another related problem in the study of sleep and dreams is the biological time clock. The duration of our day as determined by a biological clock is not 24 hours but somewhat longer and so every day we and billions of other animals have to adjust our biological clocks in order to comply with the revolution of the earth around the sun. After many years of research, the place of the clock was located near the center previously assigned to the sleep center.

Returning to the Penfield-Jasper relation at McGill, I was quite impressed by the set-up of Jasper. He was assisted by Dr. Kershman, who was about the same age as Jasper; was a very knowledgeable electroencephalographer, but was much less prepared to understand the basic phenomena from a physiological point of view than his co-workers. Incidentally, at that time Jasper was not yet a physician, and this created some personal problems. Dr. Jasper had the merit to build up and develop a method of recording EEGs during brain surgery and we, the onlookers, watched the procedure from a gallery which was glassed off. The operating room had to be specially equipped to permit EEG recordings, and there was a solemn atmosphere related to the possibility of error on the part of the electroencephalographer who, particularly at the beginning of corticography, had to direct the brain surgery which of course was irreversible. Dr. Penfield, a tall, very serious man with a charming smile, was superbly organized. He put small pieces of sterile paper with numbers written on them on the surface of the cortex, all being photographed, and then referred to the results of the stimulation of these particular points, or to the results of Jasper's EEG observations, referring back to written numbers. An epileptic patient had to disclose to Dr. Penfield his subjective experiences. Thus after the stimulation he might report, "I saw a light" or "I did not see or feel anything." At other times he would say, "I see myself with my grandmother in a garden" and then describe flowers, etc. Dr. Penfield considered these as dream-like states, which were similar to those of a seizure aura, corresponding to an area of enhanced excitability. Dr. Penfield believed that the patient's response was the result of his probing a group of nerve cells which were depositories of actual memory, primarily in the temporal cortex. He further believed that the patient's response was a normal mechanism which was revealed only in patients with highly excitable cortices. Without mentioning names, I have heard neurosurgeons say that they too stimulated the temporal cortex in individuals with epilepsy, but were never rewarded by the expression of organized dreams.

This reminded me of a story told me at the Salpetriere; prior to a lecture-demonstration of the great Charcot, who impressed the whole of Paris by the performances of his patients. For example, one of Charcot's preferred demonstrations was to hypnotize a patient in front of the audience and suggest that she disrobe, which she did. But when he told her to take a revolver and shoot him, she would not. I was told that the nurses suggested, prior to the seance, what the patient should do. Of course, Dr. Charcot never was apprised of this information.

CHAPTER 5

The Hippocampus

*"Above the valleys, the lakes, the woods,
Above the seas and the clouds
Your spirit, your soul, your moods
Fly with no fear and no doubts"*

WHY DID I CHOOSE the hippocampus as the subject of many years of my research? That is an important question. Before I left the Institute of Living I had to take the examination for a license to practice medicine. I took it both in New York and Connecticut and after I passed both, I was appointed electroencephalographer at the Hartford Hospital. The Hartford Hospital was a community hospital with a strong Protestant coloring; the other big hospitals in the city being a Catholic one, St. Francis and a smaller Jewish hospital, Mt. Sinai. Dr. Scoville, with whom I started my research, was the chief neurosurgeon; Dr. Focks was the chief neurologist and president of the staff. I was appointed Director of the EEG laboratory.

I was dismayed to find that the equipment in the laboratory was in very bad shape since no repairs were made during the war. Inasmuch as it could not be used for any research requiring precise data, I asked the Director of the hospital to give me the money to replace it. The Director listened to me and said, "We have a meeting of the Board of Directors this week. Why don't you come to the meeting and ask them for a donation? After all, they are bankers, directors of insurance companies, etc.; and have lots of money." I accepted the challenge. However, I wondered how I could explain to these bankers the challenges of EEG work? I gave it a lot of thought and finally came to a decision.

At that time there was a popular thriller out called "Sullivan's Brain."

It was a book telling the story of a wealthy man who had a lot of money hidden in an unknown place. He was flying his own plane, which crashed. The crash occurred in the near vicinity of an EEG laboratory. The scientist of the laboratory took the moribund body, brought it to the laboratory and recorded his brain waves. From this record they could identify the place where he had hidden his money. I may have distorted the details, but that was the general gist of it.

So I said, "Gentlemen, of course this is something we can not yet accomplish, but in the future it is possible that we will be able to perfuse the brain to keep it in a functional state for months, if not for years and be able to record its brain waves. At that time, each educated person will have been trained to communicate by means of his EEG. Indeed, the EEG may be considered as a succession of waves and silences; thus may be considered as a type of Morse code. So, the surviving brains could communicate with the outside world and the information could be channeled to the brain in the usual fashion." At any rate, my plan was accepted as a reasonable one by the bankers; possibly because they thought they had something to say after they were dead. They gave me the money. The Chief Neurosurgeon at this hospital was Dr. Scoville, and he of course heard of my daring proposal. He asked me in the meantime for my collaboration during his new operation on psychiatric patients.

His new operation was a variation of a therapeutic lobotomy for hopeless psychiatric patients. At the present time the idea is revolting inasmuch as we have pharmacological methods of altering behavior, but at that time, if the patient could not be improved by electric shock or insulin treatment, and if one wished to have some minimal communication with the patient, prefrontal lobotomy seemed to be the "therapeutic" method of choice. In the desire to help the patient and his family, Dr. Scoville thought that the personality changes observed with classical pre-frontal lobotomy were due to the extent and type of the surgery; he therefore decided to produce a limited undercutting of the temporal lobe which covered the hippocampus; which he expected to do much less damage to the brain physiology. The tip of the temporal lobe (the uncus), a relatively limited lesion, was undercut. To control the cortical undercutting it was important to have a neurophysiologist as a collaborator.

I administered the word association test during the surgery. The most important findings were: 1) Whenever the surgeon touched the uncus, and of course much more if electrically stimulated, the patient exhibited a frightening apnea which required several seconds for recovery. 2) The

The Hippocampus

patient stopped responding to addressed questions. This might have been related to stimulation of the speech area, but we felt it was more related to general decrease in awareness. 3) The patient manifested obvious EEG signs of seizure phenomena and clinically occasionally convulsive movements were observed. The EEG abnormalities remained local to the temporal region. Psychological testing by Milner, et. al, of the patients following surgery showed memory disturbances when the undercutting was bilateral. Incidentally, I did not wish to be part of the subsequent paper because I was not sure, as these investigators apparently were, that the memory disorders were due to the fact that memory was localized in the temporal lobe, and not due to the fact that the brain of the patient manifested obvious epileptic activity.

In short, this surgery proved not to be very helpful and was terminated. It was however significant as far as my studies were concerned since for the first time I realized the importance of the rhinencephalon in general, and the hippocampus in particular, for the generation of electric seizure patterns, and the possibility of the hippocampus to be a part of the structures sustaining awareness.

Soon after this episode I was contacted by a talented young neurophysiologist from Switzerland, Dr. Konrad Akert, who had heard about my discovery of "microsleep" and he wished to come to my laboratory to continue his studies of sleep with me. I preferred, however, to work with the hippocampus and suggested that before we tested sleep we should do some experiments on the guinea pig's hippocampus. Dr. Akert agreed and he utilized needle electrodes with which he easily localized the hippocampus without doing a craniectomy. And so the first problem was to compare the seizure propensity of the hippocampus with respect to other brain regions such as the motor cortex and thalamus, as well as many other regions. We demonstrated that the hippocampus was the most epileptogenic structure of the brain. This fact was suspected by previous investigators, particularly by Dr. Gibbs, but I believe that no one demonstrated this as convincingly as we were able to do.

As we were considering the next step in our experimentation, one day I remember it was almost midnight, we approached the guinea pig, which had the needle electrode of Dr. Akert connected to the EEG machine. We were struck by the presence of some strange sort of artifact, irregular oscillations of around 6/sec. In general, when one works on animals of this kind, such irregularity is not observed and we thus suspected some sort of artifact. But we could not find the artifact, whenever we approached the cage the artifact became more prominent. So, we searched our persons for the source of the 6/sec artifact; but we were unable to discover anything. It was only after an hour or so of searching for the source of these oscillations that we concluded in fact, this was the brain rhythm of the guinea pig's hippocampus. It was puzzling why this pattern was exhibited when we approached the cage. If we started to talk to the guinea pig we only reinforced this artifact. Usually when one stimulates an animal, or even a human subject by an auditory, or any other input signal, the brain wave activity decreases in amplitude and increases in the frequency of output. In this case the brain wave activity slowed down and presented a regular rhythm. In brief, we had discovered one of the important physiological elements of hippocampal function. Contrary to the neocortex, the paleocortex, the old cortex, responded to the stimulation by a regular 6/sec rhythm. Our experimental results were quoted in the book of Magoun, "The Living Brain," but we did not enjoy for long the feeling of discovery of an original observation. Indeed, we found out that several years before a German physician and physiologist, found such waves resulting from any stimulus to the hippocampus of animals. We then described a fundamental characteristic of the hippocampal seizure discharge. A clonic burst is initiated with a dozen or so spikes with a frequency of around 100/sec; we called this pattern a "brush." Then a slower pattern of spikes, about 15/second, which is closer to the spikes which are recorded in the human EEG of an epileptic, together with a 10 to 30 per second band. As a matter of fact, these spikes appeared to us as waves when we recorded them with the appropriate high velocity of the recording paper. We called them "pops." To make the story even more unbelievable, and yet true, we observed that the spikes and the pops were in turn underlying very fast spikes of the order of 100 to 300/sec. We called the ensemble of all of these findings the "comb and brush" pattern. Thus, a fundamental feature of the epileptic discharge in the hippocampus is an inter-relationship between the waves, or spikes, of different frequencies. This was not so, however. After Dr. Akert returned to Switzerland the research was continued with Dr. Cadhilac from Montpelier, France. We not only confirmed and amplified our observations concerning "combs and brushes" but we found that both were sitting on a slow wave or a DC potential, just as I discovered previously in relation to the spike-and-wave pattern.

Now we have several elements: 1) A DC potential which is "supporting" the pops. 2) The pops support the spikes, and 3) these spikes are supported by another episode of spikes. The primary phenomenon of interdependence exists therefore in a most meaningful way in the epilep-

The Hippocampus

tic discharge of the hippocampus. Two years later we found that Kandel, in 1961, recorded a similar relationship using microelectrodes in a single cell of the hippocampus. Finally, and more recently such a pattern was identified as one which is related to calcium participation in the membrane discharges of cells. We also found in certain experiments "typical" three per second wave-and-spike formations. With Cadhilac, we also recorded DC potentials in one direction in the clonic phase and in the other direction during the interval of silence. There was also another pattern which we recorded which was discovered prior to us, namely, a burst suppression pattern, which is usually an expression of a pathological brain. This pattern consists of a succession of irregular waves and spikes separated by intervals of silence. Our contribution was that we found these bursts also "sitting" on a DC potential which gave us an opportunity to evaluate the significance of the DC changes. Finally, we recorded a traveling DC potential in the hippocampus of the type described by Leao, which was associated with fast spindles.

It is important to note that a gigantic step in the understanding of the hippocampal cell function was made by the methodology of Craig and Hamlin in 1957, just a few years before our experimentation. The pyramidal cell and its appendages in the CA1 sector of the hippocampus is oriented vertically, perpendicular to the surface of the hippocampus; the ensemble of this network is about 203 mm deep. A microelectrode plunged parallel to the cell permits one to record different potentials as a function of time associated with a stimulus applied to the afferent fibers of the hippocampus. Under these conditions we recorded potentials evoked by stimulating the homologous region of the opposite side. A diffuse positivity together with spike production with a maximum present at about 1 mm below the cortex. This deflection was succeeded by a negative one at around 1½ mm depth 10 msec later. Both of these deflections persisted for some 50 msc and then were replaced by a deflection of the other side in each case. When we recorded these events for a considerably longer time we found that after about 300 msec a peak of considerably slower wave could be seen. It was remarkable that this slow peak could respond to a single evoked potential resembling the element of a 3/sec wave-and-spike formation of petit mal. In other words, what appears to be pathological in fact is an exaggeration and slight amplification of a "normal" potential.

Continuing our analysis of the early evoked potential (really a spike in clinical terms) we witnessed a new phenomenon. When we repeated the stimulus at a slow rate of only two per second we found a remarkable increase of the response in time and amplitude. In other words, over 1½ seconds the pyramidal cell "remembered" the previous stimulus. Indeed, in this single experiment, one could identify the elementary process of the effect of remembering a previous stimulus. Let us be precise about it. In a simple muscle, one obtains a high amplitude contraction when one stimulates by two stimuli separated by a half second. However, in the case of the muscle, the second contraction is superimposed on the first one which persists due to the first stimulus. In the case of the hippocampus, the electrical perturbation elicited by the first stimulus completely disappears, but somehow it leaves a trace so that the second stimulus generates a greater response (a facilitated response). This is equivalent to short term memory. When we continued the stimulation at two per second, the response again progressively increased in duration and amplitude until suddenly, for no known reason, this increase stopped. Instead an almost flat record appeared for a variable length of time. Then out of nowhere, a typical epileptic pattern manifested itself with the process of combs and the whole panorama of an epileptic discharge sitting on the DC potential. That we could witness the development in slow motion of an epileptic attack; how the memory of previous events coincide with the period of a state of inhibition found before a full-blown seizure is still a secret of the hippocampus.

Incidentally, this type of conditioning of the guinea pig hippocampus later became a methodology to develop chronic epileptic animals by "kindling"; kindling being the process which was described above. This kindling animal now constitutes a model of human epilepsy.

It was intriguing to witness the genesis of the pattern of epilepsy in guinea pigs by stimulating the brain, not only with different frequencies but also with different total durations of stimulation. If one stimulates at 5/sec for only $^2/_{10}$ sec, there will be only a response to 2 stimuli, because stimulating at 5 per second for 2 seconds will result in 10 stimuli. In any case, by changing the stimulus voltage one finds the necessary intensity to produce a seizure. Therefore, in this way one can change not only the duration of each stimulus, the frequency of stimulation but also the total duration of stimulation. It was found in man during electric shock therapy that when the total duration was short it was possible to develop a loss of consciousness, more or less deep, without convulsions. However, when one stimulated for electrically induced convulsions for therapeutic purposes, with a long duration stimulus time, one could produce convulsions without any latency; convulsions sort of emerged from the stimulation. By this technique there was no way of producing petit mal. In the hippocampus of the guinea pig we took advantage of the possibility of studying this phenomenon and we indeed found that when the total time of stimulation was short, a quarter of a second for example, we could produce wave-and-spike formations; this latter was very difficult to produce with the total duration of the stimulus lasting for several seconds. We therefore concluded that petit mal epilepsy may not only be due to a specific localized stimulation, but also to a relatively brief irritation of the brain produced by a lesion, or any other factor, a notion that has not yet been recognized fully.

Our research definitely rendered untenable the previously held notion that the tonic phase of a seizure arose from one area and that the clonic phase arose from another area. Not only did it lead to an understanding that the components of the seizure are initiated in the same region; it introduced a new knowledge of the ubiquity of a single cell, such as the generation of DC potentials as well as waves and spikes. As may be remembered, at the beginning of my research I found that the DC current of a certain intensity could produce, even in peripheral nerve fibers, a repeated response. These are expressions of a so-called "rhythmic excitability" studied by Fessard and publicized by Lapicque. One can see that this mode of excitability is not only induced by the electrical current, but in fact is a common occurrence in the hippocampus, and probably any type of epileptic discharge.

We then tried to understand from all these phenomena the normal function of the hippocampus. We could confirm the classical notion that the hippocampus responds to many sensory modalities such as clicks and flashes of light. When one considers the clicks, the latencies are significantly longer than those of the evoked potentials produced in the neocortex. The response is capricious and easily suppressed by habituation. This suggests that the hippocampus has an "intelligence" whereby it decides for itself, as to what it should respond.

The reactivity of the hippocampus to external stimuli and the possibility of inducing the state of hyper- or hypo-polarization by various manipulations allowed us to investigate the relationships between the different potentials. This relationship is fundamental to all hypotheses pertaining to the importance of the brain waves as modulators of incoming stimuli. We found the clearest relationship between the pulses of the DC potentials during suppression episodes. We showed that there was a remarkable potentiation of the auditory spikes at the time of the DC potential oscillation underlying the bursts. A similar relationship between the DC potential and the activity of the hippocampus to external stimuli was found during the moving DC waves (Leao). The conclusion was unmistakable that there was a definite simultaneous relationship between cortical potentials and evoked potentials.

We conclude from this action, that inter-relationship between different rhythms is of great importance; not only for normal activity, but also for the abnormal, particularly for the epileptic state. We also may conclude memory may be associated with the activity occurring in "single" evoked potential patterns, and that the hippocampus may indeed be able to participate in memory processes, as is generally assumed. We know that information is channeled to the brain by complex discharges of different rhythms. And yet, we know that in the final analysis, these rhythms can be retained by the tissues via local chemical changes that support long term memory. How these rhythms are related to the formation of specific matter or specific molecules, is closing in on the secrets of nature. It is not inappropriate to remind the reader that such problems have been tormenting physicists for the greater part of this century, mainly the interrelationship between waves and particles.

With the arrival of Dr. Cadhilac in my laboratories we were extremely anxious to develop our research studies with a particular study of the hippocampus. First of all, as a clinical electroencephalographer I was interested mostly in the processes related to epilepsy and the role of the hippocampus in such actions. As we showed in the work with Akert, and which had been earlier recognized by the Gibbs', the hippocampus is a structure particularly susceptible to epileptogenic excitation. Indeed, as we know, the majority of adult epileptic phenomena belong to hippocampal dysfunction. It is not incidental that surgery of epilepsy was in practice a surgery of the temporal lobes. Clearly, in certain cases, scars in the parietal lobe, or in the motor strip will generate focal seizures, but such cases are relatively rare. On the other hand, even in the genesis of grand mal epilepsy it appeared to us that the temporal lobe played an important contributory part in the discharge activity. Indeed, the concept of Penfield regarding the necessity of a pacemaker for synchronous bilateral manifestations of convulsive disorders to be in the "center" (centrencephalic) of the brain was quite defeated by our observations in experimental hippocampal seizures in which the excitation spread so rapidly to homologous contralateral regions of the brain through the remarkably pervasive hippocampal commissural pathways. The patients who were operated on by Dr. Scoville generally presented with grand mal seizures and it appeared erroneous to us that grand mal seizures were primarily in the domain of a hypothetical centrencephalic system. As to

The Hippocampus

petit mal, the classical pykno epilepsy and its relation to hippocampal dysfunction, it was difficult to make a definite judgment from our data. Yet we felt that certain combinations of the duration of the individual stimuli and the total time of stimulation might bring about a loss of consciousness, without the gross motor manifestations, similar to what is seen in petit mal. This fact was already brought out in our experimentation in electric shock therapy where we could quite readily dissociate the two phenomena. In fact, in later experiments we were able to produce alternate spike-and-slow waves discharges identical to those seen in human petit mal by parametric control of hippocampal stimulation; this was almost certainly devoid of any contribution from a centrencephalic detonation source.

Clinically, the importance of the hippocampus resulted from the fact that many investigators determined that this ancient part of the brain was involved in affective emotional life to a much greater degree than the more recently developed cortex. The well-known Kluver-Bucy syndrome in the primate, consisted among many other signs, of exaggerated appetite for food, hyper-sexuality and fearlessness which resulted from large temporal lobe ablations. Furthermore memory disorders, both immediate action and referential, were typical consequences of temporal/hippocampal dysfunction.

We were fascinated by the purity of the electric tracings which we obtained from the hippocampal action. Although we were using relatively gross electrodes and never actually penetrated the hippocampal nerve cells, we nevertheless recognized a repetitive functional organization. There were situations in which we could recognize a negative D.C. component during the seizure discharge which resulted from the electric stimulation. Then we would produce clonic bursts which consisted of episodic bursts in humps of prolonged potentials, on top of which would be waves of more usual duration, and on top of these there was a brush of spikes. These phenomena presented as if there were several levels of excitation, each one having a particular threshold. Once this first threshold was reached the emergence of the second occurred, etc. In these experiments we became very aware that the hippocampus received data from all the primary sensory systems. Thus, whether we stimulated the visual, auditory or somesthetic systems we would obtain evoked potentials in the hippocampus together with those in the neocortex. This demonstrated that the hippocampus might play an important function in the perception of the world around us. Others, by indirect evidence had earlier hypothesized that input data were laid down in parallel in the temporal regions (including the hippocampus) and the primary cortical receiving areas. The importance of our work was that it clearly confirmed the hypothesis of the remarkable parallel input of all sensory modalities in an explicit direct manner.

Also in the hippocampus we found the fascinating phenomenon of anticipation of repetitive stimuli, as well as the differentiation between the stimuli to which the animal had accommodated to, and those that would be novel. For example, we had one experiment in which we alternated clicks and light flashes, one flash then one click, soon the hippocampus ceased to respond to the flashes; it would only respond to the clicks. At that time we presented only light flashes one after the other; the animal recognized "the novelty" and then responded in the hippocampus, one-to-one to the light flashes. Of course, such a response pattern is consistent with a perceptual rivalry response gradient, that is, the auditory signals may have had a more prominent response in the guinea pig's brain than the visual stimuli, hence with both auditory and visual the visual response was obliterated, and with resumption of only the visual stimuli there was no longer rivalry, and therefore the one-to-one response to light became evident.

We also could produce by mechanical stimulation a wave of D.C. potential (spreading depression) and could follow its influence by the suppression of the response to somatic sensory stimulation. At this time I recalled that in some of my first recordings of the human EEG when spike and slow waves were produced, it appeared that they were sitting on a pedestal of DC potential. I was at that time unable to interpret this phenomenon, but now in relation with our new findings, it appeared that the DC potential might play an essential role in the epileptic discharge. Around this time such negative DC shifts were observed in man by Cohn, and in other vertebrates by O'Leary and his associates. Cohn believed that the negative shift resulted from the excitation as determined by the initial spike, and that the slow wave was an expression of a repolarization process; the negativity during the entire seizure persisted because of the repeated spike activity. With cessation of spikes there was a return to the natural baseline. We ourselves were not sure whether these potentials were the result of the summation of after-potentials or whether the D.C. potentials generated the convulsive discharge. Later we were intrigued by the possibility that the D.C. activity was not neuronal, but the result of neuroglia action. Indeed, it may be that the spike is an expression of excitation, while the wave may be an inhibitory phenomenon. This inhibition might be the result of recurrent

The Hippocampus

collateral axonal excitation. The reason why the hippocampus presented such beautiful examples of integration of electrical activities at the various levels is that the anatomical structures are relatively simple. The main cells, the hippocampal pyramids are arranged in vertical rows perpendicular to the surface; superior to the pyramidal cells are ramifications of the apical dendrites together with axons derived from the basket cells; at the bottom are the basal dendrites penetrating the white matter of the hippocampus. It was possible, by moving the electrode to different depths, to record a profile of the electrical potentials induced by the stimulation. It was much later that Eccles showed that this wave, the superficial following the stimulus, represented inhibition of a recurrent type, which would limit the development of excitation. Indeed, we found that during the stimulation of the hippocampus leading eventually to a development of an epileptic discharge that suddenly the activity of the pyramidal cells would cease and that a rapid activity of low amplitude would succeed to the relatively slow activity of the pyramidal cells, and it was only afterwards that the electrical expression of the seizure appeared. Thus there was a sort of silent period between the succession of ordinary evoked potentials on the one hand, and the explosive discharge itself.

Because of my interest in hippocampal discharges I was invited by the French Academy of Sciences to organize and preside over a meeting on the hippocampus in Montpelier. It was indeed a pleasure to find in this old French university city several of my friends and colleagues such as Dr. Cadhilac himself, Professor Fessard and Madame Fessard as well as representatives from every corner of the world. Among these were Herbert Jasper, who was the Honorary President. I listened with great interest to the detailed controversies after the descriptions of the evoked potentials from the hippocampus. I said in my paper that as a clinical electroencephalographer I was more interested in finding the analogy in patterns of human epilepsy and the hippocampal discharges than in the detailed localization of the processes of inhibition and excitation. This was a personal feeling of mine obviously without any prejudices. There has been a tremendous investment of human intelligence and skills in the determination of the precise anatomical and physiological relationships of excitation and of inhibition in the hippocampus and other brain structures which is leading us to a much better understanding of synaptic function. However I felt that I should not be seduced by such research, and that I should be satisfied by initiating some of the early provocative studies and return to my patients and try to find lessons which may derive from these precision studies. However, there is a fervent hope that the better understanding of synaptic processes and chemical transmission in different areas of the brain will not only generate better diagnostic procedures, which are fairly good now, but rather to a different approach in the therapy of epilepsy. I must say parenthetically that while we were doing this clinical physiological research there has been a tremendous development, mostly under the leadership of Dr. Eccles, in our understanding of the processes of excitation and inhibition at the synapses. The excitatory state, or E.S.P. and the inhibitory state, or I.S.P. produced by specific transmitters were demonstrated to be due to a wave of depolarization of the neuron in the case of excitation and repolarization in the case of inhibition. The efforts which started with Galvani, and which I personally witnessed with Wedensky, who was much interested in catelectrotonus and anelectrotonus, were finally attacked by the "new" development of amplification techniques and by the use of microelectrodes (micropipettes) to inject inside the cell with specific substances, a method which was mostly due to the ingenuity of John Eccles. These studies permitted the physiologist to quantitatively present the struggle between inhibition and excitation at that level. But it must be kept in mind, what happens on the level where an inhibition is inhibited; clinical behavior may then get very complex.

A "new" phenomenon was discovered in this work by Dr. Eccles which was designated as recurrent inhibition. When a cell discharged through its axon, it fired synchronously through collaterals to another cell whose role was to return to the original cell, to produce an inhibition. This was a sort of limited feedback which was first expounded by Renshaw at Harvard.

Despite all of these exciting disclosures presented at the meeting, I decided to remain on the fringe and participate in it only to the point of being able to follow what was going on and to be able, at a critical time, perhaps exploit the new developments for diagnosis and therapy of my patients. I felt that my experimental research on the hippocampus as well as my clinical experience with Dr. Lennox was sufficient for me now to operate my own epileptic clinic. I was dreaming about this project as I debated the problems of the physiology of the hippocampus at the meeting in Montpelier.

Incidentally, this was the first time that I became aware that a dry city could become a city of unlimited use of alcoholic beverages. At the time of our deliberations there was a period of drought in Montpelier and its surrounds, to the extent that we did not have water in the hotels, not only insufficient for washing and bathing, but even for shaving. As a result, all guests were supplied with wine for shaving (and even thirst diminution purposes).

CHAPTER 6

Nerve Conduction, Evoked Potentials and Electromyography

"To shock the innocence of maidens
To alarm them with mock desperation
To hasten the lazy thoughts' cadence
To surprise with sudden inspiration"

As noted from the time of Helmholtz, motor nerve conduction velocity was capable of being measured. One determines the time latency of response of the hand muscle to electric stimulation of the median nerve at the wrist, and then the response time stimulating the median nerve in the forearm. Suppose the first latency was 4 msec, and the second latency measured 8 msec. In order for the nerve pulse to travel from the arm to the wrist, it requires 8 minus 4 msec, or 4 msec. If the distance between the stimulated "point" in the arm and the stimulated point at the wrist was equal to 24 cm, or 240 mm, the 240/4 will give 60 mm/msec, or 60 meters/sec for the motor conduction velocity between these two points. In a great number of unpublished studies, we showed that the conduction velocity was much more rapid between the axilla and the elbow, and still more rapid between the axilla and Erb's point, or the projection of the brachial plexus and the axilla, than between the more distal points. For example, if the conduction velocity between the elbow and wrist was 60 M/sec, most likely the proximal conduction velocity would be 70 M/sec between the axilla and the arm, and 80 M/sec between Erb's point and the axilla. We have published our results concerning the differential conduction velocities observed in the lower extremities. But before we consider this in detail we will outline the method used, which was that of Dawson, or the averaging of potentials. If one stimulates the tibial nerve at the knee, obviously the nerve impulse will travel upwards toward the Cauda equina, the roots of the spinal cord located in the lower portion of the spinal canal. These roots are separated from the surface by the bone of the vertebrae, by the subcutaneous fat, fascial connective tissue and by the skin itself. But no matter how low the transmitted potential there must be some precise latency following stimulation of the nerve at the knee. The reason for the inability to record this potential is that there is accompanying noise that obliterates the potential. If we can separate these physiological potentials from the noise we can recognize these root potentials. Such a method of separating noise from the physiological potentials was precisely developed by Dawson. This was accomplished in the following way. Suppose that we have 100 "recordings" from the skin overlying the lumbar spine from the stimulus applied to the tibial nerve at the knee, we will see a number of irregular potential patterns, some going up, some going down, some large, some small. But if we know that 10 msec after delivery of the stimulus, there should be a potential always going in the same direction, always starting at precisely the same time, these will algebraically summate point by point, resulting in a potential many times larger than that corresponding to each individual potential. Other potentials may also summate, but these latter tend to be random and thereby cancel themselves out allowing the time-locked potentials to build-up their amplitudes, and thereby be recognizable as physiological phenomena. In thin individuals, and in children I could record evoked potentials, by this method, in the lumbar region through the attenuating structures with a latency range of 8 to 10 msec. This is much more rapid than in the knee to ankle segment, since the knee to lumbar region is significantly greater distance than from the knee to the ankle. With these data I wondered why electromyographers measure only the knee to ankle conduction velocities and hardly ever consider the measurements of the proximal elements. The same applies to the upper extremities. Part of the reluctance to measure the arm to the axilla may in part be due to the more painful stimulation of the axilla (brachial plexus). With my collaborators I used the summation method, stimulating the median nerve at the wrist and recording from the sub-occipital region of the head and obtained a latency of around 11 msec. If we subtract 4 msec (latency from elbow to wrist) there remains 7 msec; 7 msec for a much longer distance than one from the elbow to wrist. Later we determined the conduction velocity in a more precise way, as mentioned earlier. Now, why are the proximal conduction velocities greater than the distal ones? It would seem natural that the distal parts would have a faster conduction velocity so that there would be less delay for central processing. However, as noted earlier, this must be due to the fact that in the organization of movements, the proximal segments should move first to determine the posture of the extremity, the grasping movement constitutes the final part of the coordinated action, just as the grasping action of the foot in walking. Moreover, this organization is consistent with the discoveries of Bourguignon, who found that the chronaxies of the proximal muscles, such as the quadriceps, biceps, triceps are much greater than those of the distal muscles. And so, in order to determine the phenomenon I had to use an averaging instrument, following the general directions of Dawson, and others. The averager they used was the CAT. I was aware of the fact that the CAT was not devised for the rapid activity of the EMG, but for the slower events of the EEG. Yet in my hands it gave the first indications of the law of shorter latencies in the proximal, as compared with the distal muscles.

Next I became interested in determining the relationship between conduction velocities and aging, particularly during the first years of life. I found with my collaborators that in children the nerve potentials were of higher amplitude than in adults. And indeed I found that there was a rapid maturation of the conduction velocities in the nerves of the lower extremities, and a slightly delayed maturation of the nerves of the upper extremities. In the lower extremities the conduction velocity reached its maximum at about 4 years of age; in the upper extremities the maximum was reached at the age of about 10 years. I hypothesized that this was due to the fact that ambulatory locomotion matured much more quickly than the dexterity of the fingers.

I then became interested in sensory conduction velocity. It was obvious to me that while the conduction velocity of the sensory nerves could easily be determined following stimulation at the wrist and recording antidromically in the fingers, only the reverse would permit one to determine the conduction velocities in the sensory fibers between the wrist and elbow, elbow and axilla and axilla and Erb's point. Indeed, if one stimulates the median nerve at the wrist and records from the axilla for example, one would record the potentials, and even though one may hypothesize that the sensory fibers transmit the impulses more rapidly, this method is imprecise. A more precise method would be to record the evoked potentials from the brain stimulating successively the fingers, the wrist, the elbow and the axilla. This is what we did and found the method feasible. We were followed in this path by other investigators We were wrong, however, in one particular case of trying to determine sensory conduction velocity on the basis of producing "F" waves at the wrist and elbow. We shall return to this point in a moment. It was at Hines that we claimed, being among the first, that determination of sensory conduction velocity is of prime importance because in certain neuropathies, such as in diabetes and in alcohol abusers the sensory fibers may be involved initially. This contention was confirmed by several investigators.

Let us now turn to the study of reflexes, the H-reflexes and the F waves. The notion of reflexes was introduced by Descartes, and studied in detail by the physiologists of the last century, such as Sherrington, Wedensky, Beritoff and Uchtomski, who were among my teachers; the name reflex was retained by Pavlov for conditioned reflexes. This name represented a response of the central nervous system, through its effectors, either muscles or glands, to the stimulation. Sherrington studied the stretch reflexes, which are responses of the spindle to the muscle; the spindle sends afferent signals along large sensory nerve fibers to the spinal cord, these fibers form a monosynaptic connection with the motor neurons producing a contraction of the muscle. This monosynaptic reflex was also studied by Hoffmann, and was later designated as the H-reflex. In the clinic, stretch reflexes are elicited by the reflex hammer, but in all stretch reflexes it is the spindle that is mechanically stimulated. In the case of the H-reflex the spindles are actually short-circuited. In other words, the stimulus is applied directly to the nerve fiber which conducts the messages from the spindle and thereby activates the monosynaptic response of the muscle. It so happened that Hoffmann found the reflex in all its purity in only one muscle, the soleus of the lower extremity. This was because the threshold of the sensory fibers of the tibial nerve is lower than that of the motor fibers in this area. So when one starts stimulation a reflex response is produced with a latency of around 30 msec, then as one increases the duration of the stimulus, a muscle response with a latency of around 5 msec takes place. This phenomenon results from the fact that as one continues to increase the intensity of the stimulus one involves more and more fibers and the M response increases in amplitude while the H reflex progressively is reduced. And finally disappears. Why? The logical explanation is that when "all" the nerve fibers are activated by the stimulus, a refractory state is established by the messages propogating in both directions, antidromically and prodromically, and therefore they do not permit the influx from the spinal cord to enter the muscle, the traffic being jammed. When various other experimenters used other nerves, the small muscles of the feet, or the small muscles of the hand, they did not find the above succession of phenomena. For example, in the small muscles of the foot, stimulation of the tibial nerve would produce a motor response first and then from time to time would be able to evoke the H reflex inconsistently. This response was designated as F waves. Initially these were thought to be a monosynaptic response, but later it was found in animals, that in certain cases the maximum stimulation of the motor nerves produces a reverberating action of the motor neuron, and therefore a response conducted along the motor neuron and returned along the same pathway. The same thing happens in the small muscles of the hand. It seemed to me at that time that this phenomenon was not entirely proven and I tried to determine the sensory conduction velocity on the basis of the F wave response. I found that the conduction velocities were the same as in the motor fibers, which was entirely logical. I was wrong to do that because in each individual case I was not sure whether it was an F wave, or a true reflex. But very soon this methodology was no longer necessary because of the advent of summating computers. The method, however, of producing F waves and determining the latency in the proximal segments of the nerves remained useful, and is now being used by many investigators to test whether in pathology there is increased latency on the involved side. So even being wrong in my claim, such experimentation has contributed to clinical neurophysiology. However, if it is true that one cannot determine sensory conduction velocity on the basis of F waves their presence was significant.

As discussed previously, it is striking that in the midcomponents of the somatosensory evoked cortical potentials responses, the negative component, N45 is in the order of magnitude of one half of the latency N116, and around one quarter of the latency of N170. Thinking and rethinking these findings I remembered one day my own law of "3.3." Could it be that these two sets of data could be inter-related? In working with Giannitrapani on his remarkable book on intelligence, and confirming the "law of 3.3" by his experimental data, I could not miss the fact that there was some kind of reciprocal relationship between the latencies of the midcomponents of the SSEP and classical EEG rhythms. At that time I published my data on midcomponent latencies of the auditory EPs resulting from clicks and verbal stimuli. These latter latencies were about P45, N90; P180 and negative 280 msec (the small integral relationship is obvious). The classical latencies of the visual EPs are: 70 msec, 100 msec and 170 msec. Their reciprocals are 1000/70 = 14/sec (or the frequencies of spindles); 1000/100 = 10/sec (alpha rhythm); 1000/170 = 6/sec (theta rhythm). It appears therefore that the latencies of midcomponents of EPs in different domains are equal or proportional to the "periods" of different brain wave frequencies. Thus the SSEP midcomponents of 45, 64, 116, 132 and 239 msec correspond to 22/sec (beta); 16/sec (barbiturate intoxication); 9/sec (alpha); 7.5 and 4.2/sec (theta). This was impressive but not good enough.

Imagine a somatosensory signal arising from stimulation of the median nerve in the wrist and arriving at the cortex 20 msec later. Just as any input signal, it tends to depress or desynchronize the EEG activity as it enters the cortex. Let us assume, however, that at that precise instant of input the delta, theta, alpha and beta waves which exhibit a peak in certain preferential directions would not be suppressed. Then these surviving waves of different frequencies will emerge from the EEG tracing one half period after 20 msec post stimulus, with peaks of opposite polarity to those prevailing at the time of arrival to the cortex of the fastest signal. These half periods will occur, of course, at different real times corresponding to each individual brain rhythm. *The successive peaks thus emerging from the EEG are those of the SSEP.*

Therefore the functional significance of the major EEG rhythms and of evoked potentials may be to facilitate the effects of the arriving delayed somatosensory signal components. Our explanation of the genesis of the EP is in agreement with at least the spirit of the scanning theory of Pitts and McCulloch. It views the EEG functions as that of a time analyzer of the peripheral events. Yet the SSEP contains additional components: what about them? There are several studies suggesting that signals traveling in the non-myelinated nerve fibers elicit delayed components of evoked potentials; they must arrive at the end of the SSEP, about 350 msec post stimulus. This corresponds to the conduction velocity of non-myelinated fibers that were determined in man, in vivo. However, we do not have precise latencies because one does not know precisely when the necessary time of the slow process developing at the periphery terminates to send the necessary signal to the brain. Pending additional information as to these non-myelinated fiber signals, a remarkable observation emerged from our data. Indeed, the remaining latencies of the SSEP, as calculated per our hypothesis, are exactly those determined for the visual and the auditory EPs, namely 100, 170 and 270, with only that of 120 msec missing! Cohn demonstrated that the SS, visual and auditory potentials can be recorded in all the primary receptor areas of the cortex. Should not this fact suggest that the fastest SS signal predetermines the points in time when the successive waves of information could include the fast and delayed visual and auditory signals that might have been associated with SS ones? Our perceptions are primarily pluri-sensory. Nature or God must be wise enough to make provisions for this frequent eventuality. For example, in perceiving piano playing one sees and feels the depression of the key and "simultaneously" hears the sound. And so the described mechanism must be one of the major functions of the EEG frequency spectrum as well as that of EPs. One can then ask, what about SS stimuli that are applied proximally, or distally to the hand? One may easily agree that the more proximal stimuli applied to the arm and shoulder, are not as biologically important as those applied to the hand. On the other hand, we showed elsewhere that the trigeminal EPS mid-latency components, are also reciprocal to the major EEG rhythms. As to signals originating in the plantar region we have concluded that our hypothesis holds in this case also. Indeed, if the fastest signal arising there arrives to the cortex after 30 msec, instead of 20 all the calculated latencies will be delayed 10 msec, and this agrees within the experimental error.

Function of the EEG Spectrum in Relation to Intracerebral Activities

The general thought regarding the role of brain wave frequencies related to intrinsic brain activities is that they serve to help synchronize these activities. This is undoubtedly true as far as the intracortical processes are concerned. It is more doubtful insofar as the synchronization of activities of other intracerebral structures are concerned. It is well known that the same frequencies as those of the cortical EEG frequency spectrum are found in other structures of the brain. Hippocampal theta waves are a classical example; thalamic electrical activity is another. Recent coherence studies of the cortical and thalamic outputs show a lack of complete synchronization. Working with Dr. Scoville and his neurosurgical associates on the uncus I became aware of the extreme propensity of the rhinencephalon to generate epileptic seizures. When sometime later, as noted earlier, Dr. Akert offered me his collaboration I suggested that we experimentally review the epileptogenic propensities of major cerebral structures of the guinea pig, my preferential experimental animal. In this study we confirmed the high epileptogenic potential of the hippocampus, thus confirming the less precise evidence suggested by the Gibbs'. In addition, as noted earlier, we made the observation that each time the animal was stimulated by auditory signals a regular train of "theta" activity was elicited. We recorded the cortical activity simultaneously and observed the expected depression of the alpha frequency rhythms and a complete dissociation of their electric activity. This work was later confirmed in a study with Dr. Cadhilac.

If the primary function of the EEG spectrum is not to synchronize the electrical activities of the brain, what is its function other than those discussed previously? If one wished to decipher a message sent in Morse code a most effective means would be to arrange to transcribe the electrical series from the wires connecting the sender-receiving system. Something similar may be done in the brain by placing microelectrodes in the white matter in the vicinity of, or inside a single axon. But my approach has been somewhat different. As noted previously, I used the available measurements of published interspike intervals. One of the major discoveries of these investigators has been that even when the animal is at rest, there is continual message interchange from one brain structure to another. When the animal presents with either reflex or voluntary activity, the interspike interval shortens and approaches the refractory period of the fiber. Analyzing the interspike intervals at periods of relative rest, I found that there are "preferential frequencies" corresponding to the different structures and different animals. These have simple relationships between the different recorded frequencies. The following approximate frequencies emerged from my study: 160, 8, 60, 36 cycles/sec. Their reciprocals are multiples of an interval of 6 msec; thus, $1000/6 = 167$, $1000/12 = 83$, $1000/18 = 56$ msec, etc. Of course, such duration intervals correspond to EEG frequencies and may be a resonant coding system for the brain's recognition of environmental change.

When recorded from the motor tract, shortening of the intervals proved to be a "call for action." If the shortening is replaced by an interval of rarefaction of spikes; this appears to be related to inhibition. Thus the intracerebral language is at times loud and clear; at other times it is a mere whisper. Again, in the motor tracts a murmur translates fluctuations in the motor tone. Things are not as simple to interpret in the case of sensorial tracts, or of the association pathways. We now know that entire "pictures" of the outside world can be transmitted to the primary receiving areas of the brain. The lines and shading of the image must be translated into interspike intervals. It is still difficult to visualize; yet the process is conceivable. The shorter the interval the darker must be the shading. Yet why this constant change of darkness? And what is the meaning of the brief periods of fast spikes, expressed in the tracings by lines of fused dots? We saw that such spikes corresponded to single waves in the receiving cells representing the familiar EEG frequencies. It is tempting to think of the fast spikes as punctuating marks in the sentences: comas (the briefest), semicolons and colons (less brief) and finally periods (the longest). Or perchance the longest are the exclamation marks. It is certainly premature to think of spikes in these terms; yet it may be helpful if we consider the tracts related to "real language." The fast frequencies seem definitely related to muscle activity of the tongue; but what about the periods when the intervals are longer? Are they related to our inner language? Are these a murmur of our thoughts?

The above ideas suggested that some "injected rhythms" might be committed to memory as consecutive recordings and reveal their new presence in the resting EEG, which previously did not show them. We repeated such experiments with Ellen, associating the new rhythms with avoidance behavior, thus reinforcing their significance. To our surprise they were present in the resting EEG at the beginning of avoidance training but tended to disappear when the animals were fully trained. Was it because the "electrical memory" supported by waves, became "chemical memory" sustained by particles? Are we to repeat in neurophysiology the story of waves versus particles that so agitated modern physics? Nevertheless it seems true, that somehow the EEG frequency spectrum is intimately involved in the whole story.

During my initial recordings of evoked somatosensory evoked potentials I noticed that one did not have to place the recording electrodes in precisely the same points to obtain highly reproductive patterns. I therefore engaged the collaboration of a pre-doctoral student in pharmacology, Kim, to map out the response areas over the entire scalp. Shortly after the start of this work we noted that evoked responses could be obtained over the posterior regions of the head. However, when we reached the neck, negative potentials could be recorded with consistently shorter latencies in the range of 12–14 msec. We published these results and suggested that these potentials could arise from the upper cord, cerebellum or brainstem. Eight years later in a collaborative study with Voris, an experienced neurosurgeon we could indeed record analogous evoked potentials from the upper cord and the mesencephalon. The difference in latency between these two potentials being only about 1 or 2 msec with great individual variability, we decided we could not ascribe in each individual patient with precision, the exact origin of these potentials.

On the other hand, encouraged by these results, we developed a technique of recording dorsal root and spinal cord potentials.

Needless to say that our original observations were met with great suspicion bordering on hostility until Bickford confirmed our findings 10 years later. One knows the outcome. Shortly after our first recordings of SS brainstem potentials, auditory brain stem potentials were recorded and our findings gave rise to scores of new studies.

Turning our attention back to the mid and later components in normal individuals, and in hemiplegics, we were struck by the inter-relations of different components of these potentials. In 1975 we published our averaged somatosensory evoked potentials based on a study of 30 subjects with normal EEGs. We were struck by obvious relationships between its many negative and positive components that escaped our immediate understanding, we just mentioned the existence of such relationships. In the ensuing years we intermittently came back to this problem, but because of our involvement in "functional electrical stimulation" that we introduced into physical medicine at the same time as carrying out our studies of evoked potentials, our progress in the understanding of these complex potentials was rather slow.

And yet the problem was quite challenging. Considering only negative components the latency of an early component (32 msec) is exactly one half of the next recorded latency (64 msec). The latter is almost exactly one half of the following latency (132 msec), which is in turn one half of the finally recorded latency (239 msec). Among the positive components 45 msec latency is approximately one half of the next latency (116 msec) and almost exactly one fourth of the last recorded positive component of 170 msec.

The presence of such a simple natural relationship suggested that there is a communality between these findings and my discovery of 1938 concerning basic brain wave rhythms. This indeed was the subject of several papers in the early eighties in collaboration with Giannitrapani; yet the reason for this relationship still escaped me.

It was only in 1987 that I published an equation that suggested the reason for such relationships. The equation is as follows:

$$L(SEP) = 20 \text{ (msec)} = 1/2 * 1000/EEG \text{ (freq)}$$

Let us consider the delta brain wave rhythm: it is about 2/sec therefore:

$$L = 20 \text{ (msec)} = 250 \text{ (msec)} = 270 \text{ msec}$$

(instead of 239 msec, the actual value)

For the other S.E.P. frequencies the calculations are as follows:

3.3/sec basic rhythm:

L (SEP) = 20 + 150 = \170 (msec)   (170 msec found)

5/sec (slow theta)

L (SEP) = 20 + 100 = 120 msec   (116 msec found)

6.5/sec (fast theta)

L (SEP) = 20 + 75 = 95 msec   (too close)
                                (co-number)

10/sec (alpha)

L (SEP) = 20 + 50 = 75 msec   (64 found)

20/sec (beta)

L (SEP) = 20 + 25 = 45 msec   (45 found)

The 30 msec latency according to this equation compares to a 50/sec rhythm is too fast to be recorded with the ordinary EEG, but is close to the recently discovered 40/sec rhythm.

So after 15 years of trials we finally could find an empirical equation which is highly consistent with the data. But, what does this mean?

There is a scenario which this equation suggests: when the fastest component of the signal arrives to the cortex (20 msec post stimulation) it potentiates those brain wave rhythms which at that very instant show a high peak in a preferential direction. Thus, obviously, 1/2 period of each brain rhythm (delta, theta, alpha, beta) a series of peaks will emerge from the EEG tracing of the opposite polarity to that present at the time of arrival of the fastest signal. The equation, as well as the empirical data show that the successive latencies of these peaks are the same as that of the evoked potentials in all domains, including visual perception.

It means that a mechanism has been evolved by which different brain waves become "carrying waves" for afferent information. Indeed, it can be easily understood that the somato-sensory stimuli attending the hand most of the time are associated with visual and/or auditory information. And thus a second aspect of our hypothesis may be formulated: The delayed component of the somato-sensory signal traveling in the myelinated fibers of the peripheral nerves are best acknowledged by the brain when they reach the cortex at the moment of the corresponding peaks of the brain waves. At that moment they are mutually potentiated, thus contributing to the components of the evoked potential.

We have found in the literature a paper which strongly suggests that the hypothesis indeed is a reasonable one. Two most distinguished British neurophysiologists, Eccles and Sherrington, analyzed in the 30's the diameters of the myelinated afferent nerve fibers in the cat hind leg. They found that the diameters (and therefore conduction velocities of the corresponding fibers) are grouped together in such a way as to form a multimodal distribution. Conduction velocities related to each mode are simply interrelated, just as brain wave frequencies are, although this aspect of their findings escaped their attention. Indeed the modes they found are 15 micra, 10 micra and 5 micra.

If these had the same relationship in the median nerve in man one should expect beside a 20 msec latency, a 60 and 30 msec latency. These relations were indeed found in our empirical study. If there was a fourth mode in the distribution of the median nerve fibers, maintaining the same trend it would be 2.5 msec corresponding to 12 msec. Therefore all the early latencies of the SEP are accounted for by this hypothesis, namely 20 msec, 32 msec, 64 msec and 116 msec.

In view of the hypothetical nature of this reasoning it is indeed impressive that such coincidence is found. This would suggest that during the millennia of evolution there was a natural adjustment between the frequencies of the brain waves and the peripheral nerve latencies. What an impressive suggestion for a possible development of cerebral functional plasticity and the brain wave rhythm.

Brain Waves and the Possible Coding of the Afferent Signals

When one considers a complex stimulus with a major somatosensory component the "silent melody" that would result from this stimulus could tune a code for this stimulus in the time or frequency domain that could obviously be expressed by different amplitudes of the SEP components. It is easily seen that different stimuli will be conducted to the

Nerve Conduction, Evoked Potentials and Electromyography

brain by different combinations of the individual nerve fibers of different diameters and therefore of different conduction velocities. Therefore it is easy to imagine how these silent melodies would be different for different signals, just as it is to imagine that these silent melodies could be as easily remembered as musical memories. This may be only a preliminary model of short term (active) memory. This short term memory may be transformed into structural memory by a process of brain plasticity to be determined in the future. Such a process can be easily imagined when one visualizes how a face is immobilized in a disk or on a tape.

However, as stated above these processes are greatly facilitated when the induced rhythms or melodies are related to behavioral, biological processes. We saw that at the time preceding the jumping of our rats, intrinsically originated waves modified the melodies due only to the intrinsic stimuli. We must therefore examine these intrinsic laws.

Such a method is proposed in the law of interspike intervals. Obviously it is more difficult to evaluate the intrinsic correlations to brain activity than the extrinsic one. It may take another century to make enough progress to enlist a general conviction. Yet some beginning has been achieved. The literature is full of examples of rhythmic centrifugal mechanisms that the cortex and other large nuclei originate during behavioral reactions. Usually these accounts show successive interspike intervals in single axons. We have magnified some of these illustrations and patiently measured each successive interval in the pyramidal tract as well as in the axons corresponding to the gray nuclei and some of the association centers. We arrived at rather optimistic conclusions. There are preferential rhythms in these single axons and there is some concordance among the latter related to different animals and different structures. The rhythms around 300 c/s, 150 c/sec, 80 c/sec and 40 c/sec seem to preponderate as well as those rhythms that are identical with the frequency of the brain waves. As far as equations are concerned, it appeared that one related to interspike interval is more manageable than those related to the frequency of the discharges. The equation that I reported is as follows:

$$I = 6.6 \, N$$
$$\text{Where } N = 1/2, 1\ 2, 3, 4$$

In other words, if a receptor cell is able to discriminate a 6.6 msec duration event, it can easily evaluate the command, "counting on 5 fingers" only,

For N = 1/2	the rhythm is	300 cps
For N = 1	the rhythm is	150 cps
For N = 2	the rhythm is	75 cps
For N = 3	the rhythm is	50 cps
For N = 4	the rhythm is	30 cps

These being the frequencies of brain waves.

It is easy to imagine a mechanism by which a cell may be able to evaluate the duration of 6.6 msec and not more difficult to imagine the number of periods that can be "counted" by the cell.

The above information suggests that much of the input of intrinsic processes occurs early in the "melody" related to the evoked potentials. When it occurs late these "insertions" are transmitted by the brain wave rhythm as "carrier" frequencies. In our own study reported above, the information of the intrinsic processes related to a behavioral change occurred at about 100 msec post the onset of the evoked potential, therefore using a theta rhythm. This is also true when intrinsic processes "interest" are used to maximize attention. As is well known when intrinsic processes contribute to the evaluation of the novelty of the stimulus or the significance of it, the additional potentials occur late, about 300 msec post stimulus.

In our study of evoked potentials elicited by monosyllabic words mixed with single spikes, the intrinsic potential N300 also appeared late.

Three major papers were presented at a conference at which I reported the average SEP discussed above:

1. In the paper of John one figure clearly indicated a simple linear relationship between the latencies of different segments of the "labeled melody" (my term) related to a behavioral response. These latencies ranged between 2 and 25 msec, given the size of the rat they were within the same limits that I independently reported at the same conference. The author, however, made no comment regarding this relationship.

2. The paper of Beritoff presented a departure from the Pavlov-

ian scheme. Let us consider a dog who sees through a window a piece of meat lying on the street. According to Pavlov the perception elicited a chain of conditioned reflexes, leading to a final unconditioned one when the dog, after running to the street, eats the meat. Not so, says Beritoff. When the dog sees a piece of meat through a window an "image" of the meat developed in the brain, and with this image, constantly present, the animal behaves as if it is in immediate contact with the meat. This image driven behavior could be experimentally individualized by Beritoff. Thus when continued stimuli are presented to only one eye of a bird (pigeon), the image driven behavior could only be generalized to the opposite hemisphere as information. As noted earlier Beritoff anticipated the later work of Sperry.

3. The third paper was presented by Morrell, who showed that there are cortical cells which respond to a variety of afferent stimuli such as visual, auditory and vestibular, for example. The "melodies" (my term), usually for the presentation of one mode of stimulus (visual, for example) is different from the melody induced by another modality (auditory, for example). When both are presented several times a combined melody emerges. However, if after such repeated combined presentations only one kind of stimulus is again presented a memory of both stimuli may emerge. This appears to be similar to what occurs when visual stimuli are presented, in ordinary "photic driving," and then cessated.

In my view these experiments suggest that there is no need to look for new "connections" in the case of a conditioned response. It may appear because there are cells that record both conditioned and unconditioned stimuli. Even when the latter are absent they cause the reappearance of memory of the "combs" and melody.

It was in 1962 that we intended to map the whole surface of the scalp to represent the effect of a single SS stimulus. However, we were sidetracked by the discovery of the short duration of brain stem responses. We have now come back to brain mapping after the technique has been considerably and significantly improved. But in this work it should be ever recalled that when we record any EEG activity we are recording the time course of the electrical potential differences derived between two "excitation" points. Our most recent differences derived between two "excitation" points. Our most recent experiment carried out with brain mapping techniques shows that each signal's components appears briefly in the somatosensory area of the cortex but then very rapidly spreads to other cortical regions. Thus additional differential spatial patterns may help in coding the input information.

The Role of Brain-Wave Activity

The history of the discovery of brain waves begins with the work of Pravdich Neminsky, a physiologist from Kiev, my native city. His seminal work, which was published in Germany, was overlooked because of the start of World War I. Described in this paper were primary and secondary waves in the EEG of the dog. Berger in Germany, who read his paper, succeeded in recording such waves from the head of man, renaming Neminski's primary waves as alpha (around 10 per sec) and his secondary waves, beta (20 per sec or faster). Berger discovered that these waves were altered by perception, states of awareness and other mental activation.

Other pioneering electroencephalographers later described additional waves; deltas (about 2 per second), thetas (6 per second) and spindles (13 per second). In 1938 I showed that all these frequencies are interrelated. Indeed, I found that they were multiples of 3.3 per second waves. In fact, this latter frequency is the most constant rhythm exhibited by some epileptics. Later on, I showed that brain waves are selectively related to "active intelligence" and that those individuals whose EEG does not show an 8 per second frequency or faster are confused from organic causes.

Adrian in England discovered that brain waves may be "driven" at other frequencies in certain individuals due to intermittent photic stimulation. I in turn showed that after the flicker stops, the brain may continue to "oscillate," as if it expects to be "driven" some more. Later on, Grey Walter, also of England, showed that these expectancy waves may be easily demonstrated by an averaging methodology. Livanov, of Russia, showed that if one used a flicker stimulus for conditioning, the "driven rhythm" is indeed detected in the EEG record during the rest periods between trials. This was confirmed by John and collaborators; but we in our laboratory showed that this can be observed only while the animal is being trained, as if these "induced rhythms" were related to short term memory rather than long-term memory. Over the long haul, this phenomenon disappears.

We were able to induce driven rhythms with auditory stimulation in man and by electrical stimulation in animals. Averaging techniques were introduced by Dawson of London for what were called "evoked potentials"; these are potentials recorded from the brain to various repetitive external stimuli. It was shown that these potentials had several components; they were like a "silent melody" that the subject may remember unconsciously. If "extrinsic processes" cause evoked potentials, the latter may be modified by "intrinsic processes" during behavior. Morrell of New York demonstrated that there are cells in the cortex responding to different sensory stimuli, e.g., visual and auditory, when presented simultaneously. After several repetitions of training the cells would exhibit a response by the same pattern to either of them only, as if both stimuli were presented.

It became probable that brain potentials evoked by the original stimulus for behavior became a code, so that a repeated silent electric melody in the brain evokes this behavior after repetition. My attention then focused on the "evoked potential" components resulting from somatosensory stimuli, a stimulus effecting peripheral nerves. I found in 1975 that the latencies of the mid-components of these potentials are just as closely related to each other as the brain rhythms themselves.

In 1963 Kim and I found that the latency of the early component of the somatosensory potential was very short if recorded over the neck. We suggested that these components arose from the upper cord, mesencephalon or the cerebellum. We proved the correctness of our prediction of the upper cervical and brain stem origin with Drs. Vorhis and Uematsu during surgery (for pain). These were the first brainstem potentials described in man. We also showed that evoked potentials may be recorded in humans from the lumbar roots and spinal cord.

One must remember that a somatosensory signal effects different receptors of the skin, and each to a different degree. Each of these receptors elicits an electrical potential traveling in fibers of different diameter and at different velocities. The thinner the fiber, the lower the velocity. Thus each stimulus originates a multiple packet of information traveling at different velocities in the nerves directed to the brain, some very fast, some less fast, others still slower, and some, very slowly. To summarize: each stimulus may be characterized by different proportions of potentials traveling in fibers transmitting information to the brain with different velocities.

In 1988 we hypothesized the following scenario: when the fastest component of the *signal* arrives at the cortex at 20 msec post stimulus, it potentiates each of the different brain rhythms (deltas, thetas, alphas and betas) which at that particular instant has its highest peak of a certain polarity, negative for example. Then automatically one half of the length in time of each wave "period" later, (different for each kind of wave) peaks with the opposite polarity emerge from the EEG tracing. The period for the beta rhythm is equal to 1000/20 msec = 50 msec; the half period = 25 msec; correspondingly the half period for the alpha frequency waves = 50 msec; 83 msec for thetas; and 250 msec for deltas. If one adds 20 msec (the shortest latency time) to each of the above half periods, the latencies are: 45, 70, 103 and 270 msec respectively; we found in fact, the following average latencies: 45, 70, 116 and 239 msec in 30 normal subjects, in whom somatic sensory evoked potentials were studied.

As noted earlier, Eccles and Sherrington of Australia and England, respectively, showed in 1930 that the diameters of sensory nerve fibers can be identified on a distribution curve exhibiting successive grouping modes. I found later that there is a simple relationship between the values corresponding to these modes; namely 1 to 2 and to 3. Because of this simple relationship, I hypothesized that successive packets of information may arrive to the cortex exactly at the times of the peaks of the components of the evoked potentials which, as shown above, correspond to different brain waves, to form silent electrical melodies. Thus the different brain wave rhythms serve as "carrying" waves for the successive packets of information with a resulting mutual potentiation.

A permanent recording of the "silent melody" which represents the coding of the signal may be considered analogous to the permanent recording of music on magnetic tape. These recordings constitute the main basis for the most essential part of our mental life. Beritoff provided the most convincing experimental basis for the notion that behavior is controlled by mental images or, as I termed them mental "phantoms." In the past I provided some experiments that support this concept. Beritoff provides us with the ready-made experimental protocols to test these hypotheses if one uses modern electrophysiological methodology.

The mechanism for the permanent transcription of memories must be that of genetic coding. Experimentation in this domain has been progressing for some years in different laboratories of molecular biology.

Just as in the case of extrinsic stimuli, the intrinsic signals must be carried out by the rhythmic waves of brain activity. It turns out that the intrinsic rhythms also have simple inter-relationships. Hypotheses are formulated as to the intercellular language that permits one cell to follow directives of another cell. There have been many hypotheses formulated in the past to explain the genesis of brain waves, in particular the alpha rhythm, but I believe it is the first time that a plausible hypothesis has been formulated as to the relationship between brain rhythms and the components of the evoked potentials, these "silent melodies" to perceive and to remember.

Of course visual and auditory evoked potentials also contribute to clinical medicine. Conduction velocity in the optic nerves is one of the major parameters in the evaluation of patients with multiple sclerosis. Lesions in the brain stem are reflected in the conduction velocity determined by recording brain stem auditory potentials.

The latest components of evoked potentials, particularly those occurring at 100 and 300 msec after the auditory and somatosensory stimuli, are found to be related to attention and intelligence. They may be hopefully helpful in the diagnosis of dementia. I recently expressed my surprise to find late auditory and somatosensory potentials to have the same time course. I can explain this observation by the belief that they both are related to the fundamental brain wave rhythms, delta, theta, alpha, spindles and beta. Doctor Giannitrapani and I found that these frequencies are differentially related to different aspects of intelligence. New implications for intellectual rehabilitation are to be expected. Frequency analysis of EEG also contributes to this task.

Brain Recordings During Voluntary Movements

A problem revealed by my old study of "mental attitude" suggested that one must be able to record from the brain electrical activity corresponding to movements. Indeed, the technique of recording central evoked potentials permits one to record what happens before a certain point in time.

Let us consider a subject directed to flex a finger and/or tap a switch. The technique permits one to record not only immediately before the movement but also some time before and after. Let me first describe my results and then discuss what unfortunately I missed because of the technique I used. In 1965 I was told there was a paper in German describing brain electrical potentials and not being aware of the author of the publication I decided to make my own recording. However, the movement I studied was somewhat complicated. I asked my subjects to move their index finger toward the thumb and after a brief current move the index back. They had to do it spontaneously, randomly, but with intervals between movements of at least two seconds. Although my recordings published in 1966 were quite complex, I clearly identified a negative potential over 400 msec preceding the initiation of the movement. I therefore was satisfied that indeed a formation of the image of the movement could be electrically recorded during an appreciable period of time, a finding that I clearly spelled out in my publication. When I finally read the initial pages of the German paper I realized that these authors did describe the same phenomenon, but since they used a DC amplifier (and I did not) they identified two different phenomena preceding the voluntary current of a finger movement. An initial negative deviation lasts about one second only to suddenly be followed by a more pronounced negative potential. Apparently without using a DC amplifier I missed the preliminary one second of slowly increasing negative potential and was able only to record rapidly increasing potentials of 400 msec duration. The authors who discovered these potentials before I did, did not however mention a relationship to a "mental image." It will take additional experiments to either prove or disprove the nature of this potential. I recently edited a paper where the authors used the most modern techniques of brain mapping and contributed to the description of these events. Yet they did not mention the possible psychological nature of this preliminary potential.

Assuming that I am right, a methodology then can be found to describe electrical expression of mental image. Unfortunately in this particular case it is impossible to differentiate the mental image of the movement to perform from the expectation of the movement to perform, expectation which is known to be helped by evoked potentials.

In a study in collaboration with Fried, I modified the set-up of Grey Walter's experiments with expectancy waves. His technique is as follows: A signal is produced (click) and then one second later another signal is emitted (flick). At that time the subject is supposed to press a switch with his finger. As one repeats the paradigm a steady negative potential appears in the cortex between these two signals. This steady potential expresses expectancy. But one may also say that it expresses an "image" of the 2nd signal to come. Unfortunately with this kind of paradigm the subject tends to become drowsy and falls asleep. I changed the technique so that the second stimulus was electric shock applied to the finger on the switch. In some cases the subject was instructed to push the switch, in others not to.

Analysis of the date showed that there was a tendency to increase the negativity between the two stimuli when the subject had to push the

Nerve Conduction, Evoked Potentials and Electromyography

switch and therefore produce a movement. However, the difference was minimal. We concluded that Walter's observation and ours were fundamentally the same. Both express expectancy yet both may be a manifestation of a mental image of the stimulus or the movement to come. Further research may decide which.

ELECTROMYOGRAPHY

Electromyography has been used to study the effects of muscle relaxants and spasmolytic drugs. Although such drugs are used in cases of diseases involving the mechanisms of voluntary and reflex action (upper motor neuron diseases) a brief survey of applications of electromyography may also include its description in relation to conditions affecting the nerves supplying the muscle (lower motor neuron disease), those involving the muscle fibers primarily such as muscular dystrophy and myositis and finally those affecting motor end-plate interposed between nerve and muscle fibers and controlling the mechanisms of transmission of nerve impulses to the muscles, with a resulting muscle contraction.

In normal individuals, muscles at rest are electrically "silent." This means that they do not exhibit appreciable myoelectric potentials which, when the muscles are active, follow the transmission of a nerve impulse to an activated muscle fiber and precede the onset of the muscle contraction. During voluntarily, reflexly or electrically induced neuromuscular activity, myoelectric potentials are of relatively brief duration (about 5 to 12 thousands of a second or millisecond) and are of low amplitude, on the order of a few thousands of a volt or millivolts. Because of such low amplitude they may be recorded only after amplification which is a main function of the electromyograph. Because of the brief duration they are usually recorded by a cathode-ray oscillograph, although in cases of voluntary contractions, suitable electromechanical recorders may also be used.

Motor Unit

Myoelectric potentials are usually recorded with the help of needle electrodes inserted into the muscle (localizing electrodes). Such electrodes are needed where recording of single motor unit potentials are desired. A motor unit is made of a number of individual muscle fibers, from a few to several hundred, depending on the muscle explored, all innervated by the same single nerve fiber. In view of this, voluntary or reflex activation of all the fibers contributing to a motor unit, occurs almost simultaneously. Some differences in timing of individual muscle fibers do occur for the following reasons. These individual muscle fibers are reached by intramuscular nerve twigs of different lengths. These fine nerve twigs are branches of the same main single nerve fiber which supplies the entire motor unit. Each of these twigs reaches the motor end-plate of a particular motor fiber after traveling within the muscle for a variable distance. The end-plate of some muscle fibers may be located closer to the point where the main nerve fiber is branching off than the end-plates of some other muscle fibers. Since the conduction of these individual nerve twigs is not instantaneous, the longer the twig, the more delayed the arrival of the nerve impulse to an individual end-plate and consequently the more delayed the emergence of the corresponding individual muscle fiber potentials. Conversely, the recording needle electrode is not reached simultaneously by the myoelectric potentials originating in different muscle fibers belonging to the same motor unit. Indeed, a muscle fiber does not generate electric potentials simultaneously all over its surface. The electric potential arises in the vicinity of the motor end-plate and travels along the rest of the muscle fiber with a velocity at least ten times slower than that of the nerve impulse (about 4 meters a second instead of 40–60 meters a second). The recording needle electrode may be located near the level of the end-plate in the case of some muscle fibers, and far from this level in other muscle fibers belonging to the same motor unit. The needle electrode will, therefore, record muscle fiber potentials after variable delays resulting from the traveling of muscle potentials within each muscle fiber.

Thus the motor unit potential is a summated potential made of individual muscle fiber low amplitude brief "spikes," each lasting only one to two milliseconds. Since these spikes reach the recording electrodes after variable delays, the total duration of the summated motor unit potential, made of superimposed spikes dispersed in time, is significantly longer (5 to 12 milliseconds). It will be seen that when muscle fibers become denervated, as a result of a lower motorneurone disease, for instance nerve injury or when a part of the motor unit is destroyed as in muscle dystrophy, very brief electric spikes, sometimes of very low amplitude are recorded at rest (fibrillation) or during activity (dystrophic potentials).

When muscle contractions become stronger two phenomena are observed: one, the rate of electric potentials in each motor unit increases to about 50 c/sec; and two, the number of motor units activated by voluntary or reflex action also increases. As a result of this dual mechanism it becomes increasingly difficult to recognize individual motor unit potentials, as each potential interferes with the other. Thus a mesh of interfering potentials is recorded during intense voluntary or reflex activity (interference pattern).

Characteristics of Voluntary and Reflex Contractions

When global muscle activity is studied, needle electrodes may not always be used. When skin electrodes are used instead (surface or diffuse electrodes), interference patterns become still more complex as the activity of a greater portion of the muscle is recorded by such electrodes. Skin electrodes are not used when individual motor unit or muscle fiber potentials are to be recorded.

Voluntary Contractions

When a voluntary contraction related to an habitual motor act is performed, only muscles involved in this act exhibit electrical activity. The muscles antagonistic to those which are activated the most are usually electrically silent, as well as more distant muscles, unless they are involved in normal postural adjustments of the body during such motor acts. For example, when the trunk is bent to one side and this position is maintained, the muscles of the spinal column which are inserted on the side of the convexity of the spine, resulting from its bending, are contracted in order to counterbalance the effect of gravity. The muscles on the side of the concavity of the spine are relaxed.

When a muscle is suddenly stretched, it resists by an active contraction associated with electrical activity. This is called "stretch reflex." Tendon jerks (knee and ankle) are manifestations of such a reflex. It is assumed that percussion of a tendon is associated with a brief elongation of the muscle with a resulting stretch. The reflex which is induced in this fashion is a spinal cord reflex. It originates in the intramuscular receptors called "muscle spindles." When such a spindle is suddenly elongated, some of its structures generate a nerve impulse which travels toward the spinal cord. When the latter is reached its motoneurons respond by dispatching in turn another nerve impulse to the muscle where the stretch was produced so that the muscle is made to contract. Because it contracts, its elongation by the stretch is counterbalanced, the spindle becomes relaxed and therefore ceases to send reflexogenic messages to the spinal cord. Since only two elements of the spinal cord may be sufficient to carry out this reflex (one sensory and one motor) with only one single contact (synapse) between them, this reflex is called monosynaptic. There are more complex spinal cord reflexes involving many synapses called "polysynaptic reflexes."

Instead of the percussion of the muscle tendon, an electrical stimulus applied to a nerve supplying the muscle may bypass the spindle and yet elicit the same reflex. Such a stimulus will not only elicit a contraction of the muscle directly by originating an impulse traveling toward it by a short and brief pathway, it will also activate it reflexly by sending an impulse traveling along a sensory nerve fiber toward the spinal cord by a long pathway. Indeed, the spinal cord will reflex the stimulus toward the same muscle, this time along a motor nerve fiber. Thus muscle twitch resulting from such stimulation will be associated with two electrical potentials; one arriving shortly after the electric stimulus was applied to the nerve (usually no longer than 10 milliseconds) and the other after a longer latency time, about 30 milliseconds. The presence of two electrical potentials in the muscle instead of one is obviously not revealed by a simple inspection or recording of the twitch. It may be revealed only by electromyography.

This electrically induced reflex is called "H-reflex" because it was discovered by Professor Hoffmann. H-reflex is easily recorded in some large muscles of the leg such as the soleus or gastrocnemius involved in standing posture. It is elicited there by an electrical stimulus applied to the tibial nerve behind the knee. H-reflexes are more difficult to elicit in other muscles of the body, particularly in the small muscles of the foot and the hand. When it is successfully elicited in such small muscles, it is called the F-reflex (from foot reflex). The nature of some F contractions is a matter of controversy at the present time, but this controversy does not affect the present discussion. Indeed, it will be seen that F-reflexes become much easier to elicit in spastic muscles.

A monosynaptic reflex, be it a tendon jerk or H or F reflex, is followed by a period of electrical silence (silent period). During this period, the activity of the spinal cord motor center (motoneurons) is inhibited. A silent period is very easy to elicit in practically every muscle of the body. For this purpose the subject is asked to contract the muscle voluntarily; then a single electric stimulus is applied to the nerve supplying this muscle. This stimulus elicits one, a direct muscle contraction superimposed over the voluntary contraction; and two, an H-reflex (in some cases) and three, a suspension of voluntary electrical activity. This suspension of voluntary activity is of relatively short duration (about 1 to 2 tenths of a second) and therefore, the subject is completely unaware of this brief relaxation of the contraction. Electromyography is the best method to detect this phenomena.

Although the silent period is a very obvious and prominent phenomenon, its duration for a given muscle in a given subject is variable. It decreases with the increase of voluntary effort and in case of an extraordinarily powerful contraction, it may be completely obliterated. Finally, it must be noted that a normal individual may instantaneously relax his muscles no matter how intense the preceding voluntary effort.

In lower motor neuron disease, the following events should be considered.

Fibrillations

When the reaction of degeneration is total and therefore the nerve endings disappear in the corresponding muscle fibers, a single nerve fiber is no longer able to command, and therefore to integrate the activity of scores of muscle fibers which previously belonged to the same motor unit. On the other hand, the individual muscle fibers which are deprived of their innervation become oversensitive to all sorts of biochemical and mechanical stimuli. As a result of this over sensitivity, contrary to a normal muscle, a denervated one exhibits apparently spontaneous activity. Since this activity in individual muscle fibers is no longer synchronized by the unifying influence of the missing nerve impulse the recorded potentials will rarely be seen as a result of a statistical superimposition of single muscle fiber potentials (each being of 1 to 2 msec. duration).

Thus only brief (1 to 3 ms) electrical spikes which may be of very low amplitude spontaneously appear in denervated muscles. This activity is called fibrillation. It usually appears about three weeks after the nerve injury. When the subject attempts to contract a totally denervated muscle, no motor unit potentials occur. In a case of partial reaction of denervation, fibrillations also occur but to a more limited extent, being confined to the denervated muscle fibers only. In such cases motor unit potentials induced by either voluntary, reflex, or electrical activation do occur in the parts of the muscle that remain innervated.

I continued at the time of tenure at the Veterans Hospital of Connecticut to try and induce my friend, Dr. Offner, to build me a proper stimulator but he was too engaged with other pressing duties. So, during this time I submitted electromyography to a more scrupulous analysis. I made an observation with one of my Residents that the EMG sign of denervation, which is associated with fibrillations, was not necessarily associated with changes in chronaxie and that chronaxie changes may occur without the presence of EMG signs of denervation. In this work I pleaded for simultaneous use of both classical electrodiagnosis and EMG.

Positive Sharp Waves

In addition to fibrillations, a denervated muscle generates waves of peculiar configuration called "positive sharp waves." They exhibit a sharp onset and a progressive decay followed by a slow phase of opposite polarity. The duration of these waves may be of 10 to 50 ms. They are thought to be due to muscle fiber injury.

Poor Interference Pattern

When a patient with advanced partial denervation is asked to maximally contract a muscle, relatively high amplitude individual motor unit potentials are recorded instead of a complex mesh of interference pattern. It is then said that the muscle exhibits a poor interference pattern.

Potentials of Complex Configuration

Motor unit potentials show a two phase pattern, one positive and one negative. However, in some cases several phases are recorded. When the number of complete oscillations is above three, they are called polyphasic or multiphasic potentials. A relatively low percentage of these potentials is seen in normal muscles and their increase may permit one to follow the process of reinnervation.

Fasciculations

When the cells of the spinal cord (motoneurons) which originate motor nerve fibers are involved, spontaneous contractions of limited bundles of muscle fibers may be observed or recorded. They usually occur with lack of regularity. In contradistinction with fibrillations, which are in such cases recorded at the same time, fasciculations are visible on inspection and are able to move the overlying skin. Occasionally, fasciculations are observed in individuals without spinal cord disease. In such cases, fasciculations occur without concomitant fibrillations.

Muscular Dystrophy

In muscular dystrophy, the motor unit is affected but not necessarily destroyed in its entirety. Thus many motor units continue to respond to nerve activation. The partial destruction of the motor unit will result in lower amplitude motor unit potentials of a very brief duration and/or complex configuration. The interference pattern is still recorded despite a considerably weakened contraction, but it is often made of brief, complex and low amplitude potentials. In some cases a few fibrillations are also observed. Analogous patterns may be seen in "myositis." In such conditions of obscure origin, the muscle exhibits characteristic discharges with usually brief potentials of progressively increasing and decreasing spikes. This produces, through a loud speaker of the EMG machine, the characteristic sound of a "dive bomber."

Diseases of Motor End Plate

Myasthenia gravis is characterized by a decrease in the amplitude of motor unit potentials elicited by electrical stimuli, following intense muscle contraction. This post-tetanic exhaustion follows a brief phase of increase of motor unit potential amplitude (post-tetanic fasciculation).

Spasms

In case of a muscle spasm resulting, for example, from a protecting mechanism caused by pain (such as lumbago) muscles which are usually electrically silent at rest become the site of intense electrical activity. Even when the intensity of the spasm is not high enough to cause electrical activity at rest, a movement which usually does not involve a muscle may produce its contraction and therefore result in electrical activity which may be recorded and quantified. Because of a high level of electrical activity, a silent period following a stimulus of the muscle may be of shorter duration than when induced under the same conditions in a normal individual.

Spasticity

In a normal individual tendon, tendon jerks and H reflexes are relatively inhibited. They may be increased, particularly the tendon jerks, by an additional effort exerted by distant muscles. For example, the knee and ankle jerks are increased when the subject is asked to make a fist (Jendrassik maneuver). In a patient with spasticity following brain or spinal cord lesions, these reflexes are disinhibited. This results in an increased amplitude of tendon jerks, the appearance of repetitive H reflexes and an increased easiness with which these reflexes may be elicited in the muscles of the upper extremities.

One of the classical manifestations of spasticity is the appearance of a "clonus." This is produced by a forced stretch maintained by the physician. A brief stretch, for instance that applied to the Achilles tendon, elicits a stretch reflex, which in turn, induces a silent period characterized by relaxation of the muscle. However, as the muscle relaxes and is elongated by the continually applied stretch, the latter elicits a second reflex, followed by a second silent period, etc. Thus successive contraction and relaxation of the muscle may persist for a prolonged period of time resulting in a sustained clonus.

Electrical activity induced by spasticity may be revealed in a number of other ways. As stated above a normal individual may discontinue on command his muscle activity together with the associated electromyographic pattern no matter how high the preceding voluntary effort is. This is not so in the cases of spasticity. A spastic patient who attempts to stop his voluntary effort will exhibit a short silent period followed by a more or less prolonged after discharge of electric potentials. As mentioned, a normal muscle at rest is electrically silent. A spastic muscle is also electrically silent provided that the muscle is not stretched. The weight of the corresponding segment of the extremity is sufficient to elicit prolonged stretch reflex electric activity in a spastic muscle. This electric activity is considerably increased if there is an additional rapid stretch of the muscle. If the stretch is applied very slowly, stretch reflex activity is less pronounced in a spastic muscle. These characteristics differentiate a spastic muscle such as in a case of a stroke patient with hemiplegia, from a muscle in a case of Parkinsonian rigidity. The latter exhibits a continuous electrical activity even without any stretch. On the other hand, even a slow stretch increases greatly the electrical activity in a Parkinsonian muscle.

There is another difference between spasticity and rigidity. In spas-

Nerve Conduction, Evoked Potentials and Electromyography 129

ticity, electrical activity is increased under stretch in some muscles but not in their antagonists. For example, in biceps more than in triceps, in flexors of the wrist and fingers more than in the extensors. In rigidity, both extensors and flexors exhibit a marked increase of electrical activity under stretch.

Finally, spasticity in cases of spinal cord injury has additional characteristics which are not found in spasticity of a hemiplegic. It is true that in the latter spasticity is increased by all sorts of factors, particularly emotional excitability of the patient. However, this increase of spasticity is not manifested by sudden violent and often painful jerks such as those found in patients with spinal cord injuries. These spasms express very intense polysynaptic reflexes.

Quantification of EMG

In order to quantify electrical activity several procedures may be used. One, the simplest procedure is to count the number of potentials occurring during a given time interval. Obviously this becomes very difficult when the contraction is intense and interference pattern is very pronounced. A more or less subjective description may be used as a result of inspection of such records, such as different degrees of the interference pattern. Two, electrical activity may be integrated. For example, each motor unit potential may contribute to a charge of a condenser and the total charge is evaluated by direct current recording. It has been shown that such integrated current is proportional to the intensity of a muscle contraction, provided that the measurement is made under the same conditions, with the same electrodes applied to the same area of the skin overlying the muscle. Three, there are various automatic electric pulse counters which contribute to the evaluation of the muscle electrogenesis.

Drug Action

An active muscle relaxant of spasmolytic drug decreases the amount of voluntary and reflex activity and increases the duration of a silent period in the direction of increasing normality. Thus, electrical activity decreases at rest; it decreases or disappears during stretch; it decreases or disappears in those muscles which are usually not involved in an habitual voluntary act. In a patient with exaggerated tendon or H reflexes these drugs depress the electrical activity related to these reflexes.

A sustained clonus becomes a non-sustained one or is abolished altogether. In a patient in which the silent period is obliterated, it reappears once again. In a patient with after discharge following a voluntary muscle contraction, the after discharge is either shortened or completely abolished.

The above described methods of quantification of muscle potentials should be used in order to evaluate these changes.

Two additional considerations should be kept in mind related to the price one has to pay in order to obtain muscle relaxation. One, a muscle relaxing drug may reduce not only the abnormal muscle activity but also the normal one, thus decreasing the efficiency of the normal muscle activity of the patient. Two, a muscle relaxant or spasmolytic drug may induce at the same time a state of drowsiness which may interfere with the activity of daily living of the patient as well as with is occupational pursuits. There are instances when it may be important to pay the price on one or both accounts in order to reduce the abnormal muscle activity. It is obvious however, that a drug which will selectively reduce abnormal muscle activity without cutting down normal muscle stretch and without making the patient drowsy is more desirable. At any rate, in evaluating the drug action, it is essential to know the price one pays for its effects. EMG contributes to such an evaluation by showing the degree of reduction of the normal muscle activity just as EEG contributes to the evaluation of drowsiness.

Finally, in any case of drug evaluation, one should keep in mind that abnormal muscle activity is very variable and depends upon the emotional state of the patient. It is therefore important to compare the EMG findings with and without drugs under the same emotion generating conditions. Here again a muscle relaxing drug may decrease the emotional reactivity of the patient which may be all for the better, yet the above-described battery of tests may reveal the true mechanism of the drug action.

CHAPTER 7

Human Locomotion

"A reminder of his life full and proud
It outlived all gold and stones
His call for freedom, vibrant and loud
Revived dead muscles and bones"

THE STUDIES OF human locomotion were prompted for a number of reasons. Obviously one was for the physician to understand the troubles of locomotion exhibited by some of their patients, those with stroke, those with cerebellar disorders, those with loss of afferent and efferent nerve action associated with neuropathies and those afflicted with arthritic processes of various joints.

But studying the history of gait and volitional movement research, one is struck by the observation that most of the studies resulted from sheer curiosity of scientists. As a matter of fact this curiosity was not related to locomotion in man, but to birds and insects. The flight of birds was, of course, early studied and emulated by the famous Greek scientist, Daedalus and his son Icarus; and more modernly by the French scientist Marey. Two major industries were finally developed by the aid of such studies, those of aviation and of cinematography.

Two scientific areas were particularly involved in these studies; biomechanics, the study of physical forces involved in this action, and neurophysiology inasmuch as it is a neuro-muscular action applied to our bones and other tissues that assures the essential freedom for man to walk and run. To understand these actions one must have a good idea of the processes of motor command, of feedback, of participation of different anatomical structures and of the interaction of physiological and physical factors. The Russian neurophysiologist Nicolai Bernstein, who is credited with the early introduction of cybernetic ideas, believed that the study of locomotion was crucial for the understanding of the laws of neurophysiology of voluntary movements.

My interest in the field of voluntary movements began sixty years ago in 1932 when I joined the Laboratory of the Organization of Human Work, attached to the National Research Council at the Sorbonne. In addition to walking and running, human movements such as typing, piano playing, use of tools, tennis, etc., consist of rhythmic activities. Therefore it was the nature of rhythmic movements that became the theme of my early research in neurophysiology. How one perceives the rhythm of one's movements thus came under study. This was obviously related to the following question: how does one evaluate the frequencies of one's own movements and how does one know that one motor rhythm is faster or slower than another. These were the questions that I attacked with my student Lavorsky.

The initial experimental procedure was quite simple. The subject was required to operate an uncomplicated, low inertia switch, the movements of which were recorded and the motion of the fingers or wrist were measured. Initially the subject had to follow a given rhythm of a metronome, then was allowed to operate the switch spontaneously. The task was to develop a rhythm just faster than one per second, making sure that the rate was only very slightly faster than the preceding one, then to elicit a rhythm only slightly faster until the fastest possible rhythm was achieved. Records of each experiment were duly examined and the graphs expressed the successive voluntary accelerations of the subject's differential rhythms of voluntary movements. In that no auditory cues were given, subjects were forced to rely on their kinesthetic perceptions. It was found that for each initial rhythm the one as slightly faster as possible was variable so that a distribution histogram could be drawn expressing the related probabilities. Thus for example some of the rhythms intended to be slightly faster than the preceding ones were in fact slightly slower. The statistical parameters of these distributions permitted us to determine that there were two distinct accelerations' one for the lower frequencies up to approximately 3 per second, and those of higher frequencies above 3 per second. So the intention of producing a slightly faster rhythm was purely probabilistic, the brain being obviously capable of making these determinations unknown not only to the subject but also to us. The significance of the critical value of 3/sec was totally unknown to us at that time. It was eight years later that I realized that 3 c/sec was a basic rhythm, the multiples of which were called by others, ignoring these findings, delta, theta, alpha and beta rhythms, and that this basic rhythm was abnormal and prodigiously intensified in patients with petit mal epilepsy.

The cleaning woman who watched in amazement two seemingly normal adults playing daily with a telegraph key with no apparent results except sheets of paper filled with number finally confronted us with: "What is all this for?" I confess my answer was profoundly inadequate as I did not know at that time that the laws of probability we were obtaining would lead to the discovery of a fundamental brain rhythm.

One central independent observation made at that time was that the maximum possible rhythm intended by the subject to be the most rapid succession of finger flexion was significantly faster than when the subject had to repeatedly extend the finger. The usual highest frequency of the latter maneuver was about 2 c/sec in the range of delta frequency as I learned years later. This specific discovery of antagonistic movements could be substantiated when I asked the subject to close and open their eyelids and to protrude and retract their tongues at their maximal rates. These experiments were carried out at about the time Bernstein and many other physiologists claimed that such maneuvers were guided by peripheral feedback using the intramuscular spindles or the elemental "spies" of the Golgi tendon organs.

I for my part, saw in my findings that it was the mental attitude of the subject that played a major guiding role in the organization of these voluntary movements, or what we would later call, on a cerebral plan, as to how movements should be in order to comply with our wishes. Not that I denied the important role of feedback. Indeed, in my first publication on the subject I demonstrated that while making sure that the subject executed the successive movements of extension, they were quite slow and clumsy. Yet if one asked the subject to rapidly touch a sheet of paper placed near the dorsal aspect of the finger, at once the successive movements of extension became regular and rapid. I found even then that the mental attitude to touch the paper was of decisive importance. Later I calculated that when the movements are rapid there is no time for effective feedback. Indeed, the minimal time to reach the cortex from the hand is 20 msec, 40 msec both ways, and the muscle contraction takes another 100 msec, thus if the resulting movement is increased, one is at least 140 msec behind the action. This is why the mechanism of feedback guidance could hardly be reliable for fast movements. Feedback, however, is of decisive importance during training and during slow complex movements. In the case of locomotion both the cortical plan of the movement and feedback are important, and Bernstein provided us with considerable documentation on this score. In general he claimed that the lessons derived from the study of gait might have considerable importance in the field of neurophysiology.

Movements of the upper extremity involve most challenging problems; such as how to place a grasping finger in a particular point of space. I am sure that most of my readers would say: "simple." But one must coordinate the information transmitted from the retina with the commands to the muscles of the shoulder, elbow and 15 joints of the hand. In fact, this is an incredibly difficult operation. First of all, if we had to immobilize the head and eyes in a certain direction every time we did something with our hands, our efficiency would plummet. Ask a pianist to play something relatively simple with his eyelids closed and at once you will see that this does not represent any essential difficulty for a professional musician. Watch a typist copying a text, you will observe that she is looking at the text, and not on the keys. Thus it is clear that space somehow is represented in our brain, a notice that was explicitly expressed by Bernstein, but I do not believe that even he knew how it was accomplished. I performed the following simple experiment to explain the action. Standing near a wall (say a couple of feet from it), I asked the subject to touch a hook screwed to the wall, with eyelids open. I then measured the range of abduction, flexion and extension of the shoulder, flexion of the elbow and of the wrist, as well as the position of the fingers. I then asked the same subject to reach the same hook in any other way. Although this could be achieved by all sorts of distortions, it became clear that there was only one way to execute the movement in an easy and natural way. Last, I asked the same subject to hold a hook naturally; to stand erect and try to change the joint angles. I was convinced that indeed, from any point of departure a point in space may be reached, simply with only one combination of joint angles. In other words, the space around us can be defined by the configuration of joint angles. This, of course, immeasurable simplified our task.

Some investigators assumed that it was mainly the "muscle sense" that was involved, so I made the following simple experiment. I asked the subject to reach a particular elbow joint angle repeatedly and to note the final position with the eyelids open. Then I asked that the same movements be made with the eyelids closed. The same angles were made with only slight error. Then I tetanized the biceps (electrically) which inactivated the muscle receptors; the previous angle was again reached with only a minimal increase in error. This, of course, proved that the muscle sense was not the major means of spatial orientation. Usually, in the case of precision movements the action is carried out in two steps; one to bring the finger in the vicinity of the goal and then visual control to grasp the object. Obviously when one has to place a thread in a needle visual control is a must.

All voluntary movements are goal directed with a mental representation of the goal coinciding with subconsciously (conditioned) calculated corresponding points in space. We all carry with us a spatial image of the bumpers of our car and when we move from a large to a small car the image is almost instantaneously changed in space. Furthermore, we do not "think" how far to rotate the steering wheel, that spatial image is also interpreted (automatically) in our brains. The tongue does not have muscle spindles and yet one can easily write with one's tongue in space without any preliminary training. Apparently when we learn how to write we not only learn how to place the tip of the pen on paper, we also learn the figures in space that we produce; and the brain has inherent mechanisms for the means to enlarge or diminish them proportionally. As an experiment, place a pen (with eyelids closed) between your two fingers and ask yourself, "What is the direction of the tip?" You will have no hesitation in making a guess, even if you are mistaken to a slight extent. This means that the brain has inherent mechanisms to deal with directions in space.

As is well known Hubel and Wiesel received the Nobel prize, in part, because they demonstrated that certain visual cortical cells are specialized to recognize lines inclined at various angles. There must be other cells that tell us the direction in space when we think of it. It is with this in mind that one speaks of cortical cells which reproduce the face of our grandmother when we think of her, and not one of them but many.

It is obvious that in addition to the mechanism that we use to propel ourselves on the ground we also carry a spatial image of where our foot is going to land. Remember the shock you experienced when you unconsciously miscalculated thinking of an additional step down an escalator and were rudely reminded of your error by hitting the ground!

Let us now consider the terms used in the study of the gait. Imagine the movement when you move your posterior foot forward. This point is called the *toe off*, when this moving leg is elongated we call this, *vertical leg*. Finally when it reaches the floor we call the movement, *heel strike*. This is followed by *foot flat*, and again by *toe off*. The interval between toe off and heel strike is called the *swing phase*. The interval between the heel strike to toe off, is called the *stand phase*, and a short period when both legs are fixed, is called a time of *double support*.

Now stand on the floor with both feet, one in front, the other behind the center of gravity of your body and shift the weight to the forward leg. Obviously this essential act of moving the weight from one leg to the other is of major significance for locomotion, accomplished in the double support period with minimal leg displacement. This act is performed by the hidden muscles around the hip joints. Yet this essential act will not appreciably advance you from your base of support. The forward leg could have acted by grabbing contact with the floor and pulling the body forward, however instead, the center of gravity so much lighter than the earth (Sic!) moves forward. The posterior leg pushing on the floor backwards, contributes to the displacement of the center of gravity. Both actions with the feet remaining steady on the same spot accomplish the task because of the friction between the feet and the ground. Without such friction, as in the case of a steel wire, the center of gravity would remain in the same place. It is because of friction added to the rotation of the wheel that the latter propels a moving vehicle. One frequently states that what nature could not invent was a wheel, which allegedly is a product of the human brain. Nothing can be further from the truth.

In reality in the standing phase the transfer of the center of gravity of the body from the posterior to the anterior leg is accomplished with much less effort than one may imagine. Yet in order to move the body to another base of support the swing phase is essential. While for each leg the standing phase corresponds to the rotation of the leg around the ankle joint, the swinging phase assists in rotation of the leg around the hip. This phase is much more complex than the stance phase.

If we ask any one what one does with the "posterior" leg when one moves it forward, the answer is usually, "It moves forward" until the heel strike. In fact only the thigh moves forward in the first phase of the swing, the lower leg remains behind by inertia, thus responsible for the initial knee flexion. Then curiously enough the movement of the thigh slows down in order to project the lower leg forward, also by inertia. Thus the essential mechanics of the gait is accomplished by inertia as far as the knee joint is concerned. Of course the iliopsoas expends a great deal of energy for the purpose. Unfortunately this action cannot be recorded by the EMG as the muscles are too deeply seated. Another muscle, the tibialis anticus, is however called to duty at the end of the phase in order to pick up the toes which otherwise scrape the floor and impede the gait. The largest muscle of the leg, the quadriceps, remains entirely silent during the swing. The hamstrings are only minimally involved to restrain the movement forward of the lower leg.

Human Locomotion

On the contrary, during the stance phase almost all of the muscles of the leg are involved. Obviously the gluteus maximus is used to facilitate the "oar action" of the leg and to avoid the jack-knifing of the body over the thigh. Gluteus medius is used to maintain the horizontality of the pelvis which otherwise would drop to excess by the weight of the opposite swinging leg; the quadriceps are used to prevent jack-knifing of the knee. Yet it does relax slightly immediately after the heel strike to accomplish the function of a spring. The second contraction of the tibialis anticus and finally in the second half of the stance phase with the contraction of the powerful gastrocnemius, the flexors of the toe grab the floor.

It has been said repeatedly that the gastrocnemius contracts in order to push the body forward. It is indeed the action of the muscle during running; which is not used in strict walking. However, the gastrocnemius must contract to maintain the angle of the ankle in order to keep one from falling on one's face. Indeed, when the gastrocnemius relaxes at the toe off, the body does fall forward but at that time the forward leg rescues us without fail. I have, I believe successfully, demonstrated by my EMG records the falsity of the gastrocnemius push off in strict gait.

At the beginning of the study of gait, as developed by Bernstein, photography was utilized. Small electric bulbs were attached to the costume worn by the subject at the center of gravity, at the hip, knee and ankle joints. The lights flashed as the subject walked and the trajectories were photographed by still cameras. The records consisted of points of light separated by intervals proportional to the speed of motion at a particular joint, as well as the body. From these points the velocity and accelerations could be laboriously calculated. One day I heard that the well known Parisian Engineering Company of Baudoin had developed accelerographs to test airplanes in flight. This instrument used piezo-electric principles discovered by the remarkable physicist, Pierre Curie, the husband of Marie Curie. By this method the laborious calculations could be obviated. The principle of this new methodology was as follows: Suppose you sat in a car and a pressure sensitive device was placed at your back. If the car suddenly accelerated the inertia of your body would increase the pressure, if the car decelerated the pressure would proportionally decrease. My plan was to attach these crystals to a weight to apply initial pressure, and then get the subject to walk. One can imagine the excitement of my co-workers when I showed that the record I obtained within seconds, was similar to the Bernstein records that required days to calculate. Later the quartz crystal instruments were replaced by strain gauges. But EMG replaced these techniques by its ease of operation and clinical usefulness.

Nevertheless, let me point out with some pride that the piezoelectric method was the first successful attempt to study gait without photography, but as it so often occurs in the history of scientific discovery, almost simultaneously American investigators developed the study of gait by using a recording platform. They were able to measure some essential mechanical characteristics of the gait. Their methodology was much more expensive than mine; and it was not portable. It analyzed one step at a time and it did not immediately generate a record. However, I must admit that the platform method benefited from superior marketing that I was not capable of duplicating.

With my accelerographs the following normal tracings were obtained: At the beginning of hip flexion the angular accelerogram shows an initial inflection. It reaches its maximum at the time when the thigh movement slows down. From this point on down to the baseline the angular accelerogram is coincident with the movement of the thigh forward but at the crossing of the baseline, the standing leg becomes vertical. The end of the descent indicates, of course, the heel strike and the short sequel corresponds to the period of flattening of the foot. In addition, we established that exactly at the upper tip of the angular accelerogram the dorsiflexion of the foot starts. The maximal dorsiflexion coincides with the crossing of the angular accelerogram at the baseline.

We confirmed the significance of the double peaks of the vertical accelerogram formulated by Bernstein, bringing additional details. Indeed, the beginning of the final peak is due to the contraction of the gastrocnemius on one side while the beginning of the second peak corresponds to the contraction of the gluteus maximus on the opposite side. We established this taking into consideration the time interval between the EMG and the actual mechanical result of the muscle activity. The trough between these two peaks is due to the passive flexion of the knee, reproducing the slow absorber mechanism.

Thus, so far we could identify the activity of four important muscles without recording the EMG, just by observing three accelerator tracings, which are more easily recorded than the EMG.

1. Activity of the *Iliopsoas*: Beginning of the angular accelerogram; the decreased action of the muscle corresponds to the peak of the accelerogram; its cessation: to the end of the accelerogram.

2. The activity of the *gluteus maximus*: By the beginning of the second hump of the vertical accelerogram; the end of its activity corresponds to the beginning of the iliopsoas.

3. The contraction of the *gastrocnemius*: The beginning of the first hump of the vertical accelerogram. The end: the beginning of the angular accelerogram.

4. The beginning of the contraction of the *quadriceps*: The end of the trough of the vertical accelerogram. The end is obviously at the beginning of the angular accelerogram.

5. The beginning of the contraction of the *tibialis anticus*: Corresponds to the tip of the accelerogram.

We could also localize one of two contractions of the hamstrings and the second contraction of the tibialis anticus.

If we use another accelerograph looking sidewise we could identify the beginning of the contraction of the gluteus medius, as well. Moreover, the following landmarks can be established from the accelerograph recording.

1. Toe-off; Beginning of the angular accelerogram.

2. Vertical standing leg and foot flat: Crossing the baseline by the angular accelerogram.

3. Heel strike: End of the vertical accelerogram.

4. Passive flexion of the knee: Through the vertical accelerogram.

5. End of the first passive flexion of the knee: Top of the vertical accelerogram.

6. Beginning dorsi-flexion of the foot: Same point.

Thus we got from our methodology more than we bargained for: the possibility of getting by without electrogoniograms of the hip, knee and even ankle, at least in normal subjects.

The goniogram of the hip is a simple monophasic curve. Its beginning corresponds to the contraction of the psoas; its ending the contraction of the gluteus maximus. The goniogram of the knee is more complex: initial flexion (passive) followed by extension (passive) followed by the first extension, preceded by knee flexion (passive), followed by extension (active). The goniogram of the ankle, is even more complex: initial passive plantar flexion terminating by the beginning of active dorsiflexion, followed by the second passive plantar flexion slowed down by the extension contraction (second) of the tibialis anticus followed by the second (passive) dorsiflexion, followed by the third (active) plantar flexion. Beginning of the first plantar flexion, starts with the beginning of the angular goniogram; the beginning of the first dorsiflexion starts with the tip of the angular accelerogram; the beginning of the second plantar flexion starts with the end of the angular accelerogram, and the beginning of the passive dorsiflexion starts with the beginning of the angular accelerogram on the opposite side, the beginning of the active plantar dorsiflexion starts before the first vertical hump of the vertical accelerogram.

Let us discuss now a completely unexpected discovery. The ankle electrogoniogram reproduces exactly the angular accelerogram of the leg segments present above the ankle. It is as if the foot were the "conductor" of the events related to the interplay of the anatomical and mechanical factors of the gait. Indeed, what we record during locomotion is the melody of electrical rhythms generated in the spinal cord, coordinated with the suprasegmental activity of the central nervous system. At the same time these nerve centers receive commands for the movement of the foot as well as the thigh and the lower leg. The command issued to the foot muscles is obviously conditioned by its interaction with the floor during locomotion. Apparently the same neural centers issue the command to the muscles of the thigh and lower leg such that the resulting angular accelerogram of the lower extremities duplicates the pattern of the movement of the foot. These patterns of command correspond to a symphony of melodies elaborated by the electrical activity of the nervous system, both the brain and the spinal. Decades will probably pass before we start to understand this wonderful interplay of anatomical, physiological and mechanical factors.

I have omitted some factors upon which previous descriptions of gait were based. Indeed, it has been stressed that the following elements are important: one, the lower the center of gravity the less work we have to do in walking. This lowering of the center of gravity is achieved by a slight flexion at the knee joints and by an increase in the distance between the feet when walking; two, an economy of the energy expended during walking may be achieved by rotation of the hips, partly insuring an advance of the forward leg; and three, the lateral inclination of the body over to the side of the moving leg somewhat lowers the center of gravity. I would like to complete this description of the normal gait by the following comment. We are accustomed to seeing people walking in the street with prominent associated arm movements as if these movements contribute to our locomotion. Yet if something is held in the arms, the walking operation does not appear impeded. It seems that this arm movement is purely an atavistic sign bearing witness to our evolutionary development from the quadruped, an observation ignored by most observers. However, it is obvious that a running man with a non-moving flail arm does appear to be clumsy.

When discussing the goal of movements we must consider the problem of evaluating space. When we move our hand to grasp an object it would obviously help if the brain was able to identify all the points of space that surround us. How does the brain know how to move the tip of our index finger to a specific point in space, imaged by the brain? In order to do so, millions of muscle fibers have to be mobilized to a variable degree with different speeds. As an example, let us go to the wall and place the tip of our index finger on a specific point of the wall and press hard enough so that this point will not be changed. Then, without changing the position of our feet or of our body, try to maintain the finger at the chosen spot on the wall without changing the shoulder, elbow and waist angles. We can easily see that when we do that to a certain limited degree, the different angles achieved do not seem to be natural. In other words, we can conclude that the simplest thing to identify a point of space is to assume a certain combination of the active joints involved. It means that we know the space that surrounds us by the combination of joint angle that would permit us to reach it easily. It is true that by using unusual joint angles we can reach the same point but this is not the usual way to identify the points of space that surround us. This is not only of theoretical interest. When we build a motorized brace for a paralyzed patient we may use this principle of reaching a combination of joint angles to reach a specific point of space. Complicated but possible to achieve. Since the movement is reaching different points of space the task of the brain is to achieve different combinations of joint angles after years of learning during infancy and early life.

What about movements beyond the space surrounding us in the sitting position? Consider walking to a door or a window or to any specific object. Obviously these movements consist in rotation of the body (mental image of corresponding angle joints) and straight walking.

How about moving a car in certain directions? Here a fundamental new mechanism emerges. It is based on the principle of substitution. The rotation of the steering wheel clockwise means turning the front of the car to the right. Yet in order to do so the brain must be aware of where the front of the car is located. Changing one car to a smaller one automatically changes our expectation of a new location of the front of the car. This notion of substitution is essential for voluntary movements. It is not only related to the results of the motor action but also to perception. Think of touching an object with a pencil. You have a perception located at the end of the pencil, although of course there are no perceptual structures there. Perception is purely imaginary deduced from that of the pencil. Close your eyes with a pencil between two fingers and ask yourself where is the tip of the pencil. You will have an image of its tip and its location, whether you are right or wrong. In other words, the brain will calculate the best it can the point of the location.

I give you these examples to illustrate the complexity of voluntary movements and of the task of the brain. It must continually imagine and calculate. We have tried to understand its ways in a complex experiment. It did reveal to us that the brain relies to a certain extent on probabilities. In our experiment the subject was told to press a key at a frequency of once per second using a metronome. Then he was told to do it faster but as little faster as possible. And so on until the subject reached his maximal frequency. Analysis of this data was quite revealing. When we plotted the percent of the increase of the frequency of the movements against the preceding rhythms it revealed an interesting finding.

Starting with a certain frequency all the slower rhythms exhibited showed a relatively constant percentage of increase. Likewise, above this critical frequency there was a percentage of increase but it was significantly smaller that the other. When we asked the subject they reported that the movements seemed to be different when they were below this critical point from how they seemed to be above it. Below this limit the movements were perceived in isolation one after the other. Above this point the movements were perceived as "rhythms," not isolated single movements. The most interesting finding (realized long after the publication of the paper) was that this point in time separating two types of performances was equal to Pi per second, the fundamental rhythm of brain waves.

Rehabilitation and Neuromuscular Disorders

I shall limit myself to problems dealing with the biological aspects of rehabilitation in neuromuscular disorders. This does not mean that I do not consider research in social and psychological adjustment of these patients as less promising or less important. Nor does it mean that I do not value the method of statistical analysis and prediction based on establishing a sort of inventory of all deficiencies and assets of patients which may be easily observable and rated. On the contrary, I think that these latter data are of great importance and may be used in the field of rehabilitation of patients with neuromuscular disorders. One may hope that the scope of such research will increase during the present and coming decades.

Among related problems, effect of age on the disability and the data emerging from the basic psychological and PM&R tests (such as activities of daily living; manual muscle and related tests: range of motion tests; pre-vocation workup; intelligence and projective tests) are to be correlated with the outcome of rehabilitation. A sound statistical study of such correlations is bound to permit professional personnel, so scarce in the field of rehabilitation, to be utilized for the most rehabilitable patients. Dr. Peszczynski and his group in Cleveland have been carrying out important research in this field.

I prefer, however, to discuss more specific biological methods and aspects of rehabilitation, depending upon the progress of laboratory techniques such as electromyography, electroencephalography, accelerography, high speed photography and neuropharmacology. The fundamental approach consists in viewing rehabilitation as an intensification of a biological adaptation acquired in the course of the phylogenetic development of different species, including man, as reaction to disability resulting from system or organ deprivation. Take an example of a hemiplegic patient. His involved leg is longer than normal because of a foot drop. So quite naturally, without any training, he will circumduct during attempts to walk. We have all observed quadriplegics using both of their wrists for the movement of prehension, a procedure which some of them prefer to wearing complicated splints. This ability to readjust to system or organ deprivation leads to amazingly effective corrective techniques in children. We should all learn from our patients the natural procedure of adaptation to physical disabilities. Some procedures may be taught to the patient, both as a result of accumulated experience and the result of evolvement of rational rehabilitation techniques based on our knowledge of anatomy, physiology and psychology of man. Obviously, different schools may express controversial opinions from which the best methods will finally emerge. For example, in a patient with an extensor tendon transplated to the tendon of the flexor of the same joint, one may ask oneself which should be the better technique: Should one teach the patient to deliberately try to extend the joint each time he wishes to flex it? Or should one suggest to the patient to try to flex the joint until a proper new coordination evolves as a result of repeated frustrations, and occasional successes? Only experience will allow one to choose the proper method in specific cases; yet the therapist should be prepared to offer concrete choices of action to the patient based on some rational basis. The necessity of a systematic training of the patient given a prosthesis, or a functional splint, or a self-help device is evident. Yet, here again one should analyze first the biological mechanisms of compensation; otherwise one may have the bitter experience of seeing the patients rejecting, one after the other, the devices which a priori seem to be ideal for them. Obviously, one cannot expect to understand the natural adaptive reaction of the patient to a disability if one does not have real knowledge of the fundamental processes of adaptation of a normal individual to the demands of the environment. One cannot expect to understand the pattern of an abnormal movement if one does not have present in one's own mind the principles of carrying out a normal voluntary movement. And so, the most important program for the diseases of the neuromuscular apparatus must come from our better understanding of the psychophysiology of normal complex movements, the principles of their coordination and training.

A major new insight into the genesis of movement pattern came with a better realization of the importance of sensory components of the coordination of movements. Obviously it was known to physicians since the last century that patients with tabes dorsalis cannot stand or ambulate with the eyelids closed, as they do not have the sense of "position" of their extremities. The famous English physiologist, Sherrington, made classical studies at the turn of this century, showing the importance of the intramuscular receptors for the proper adjustment of the muscle contraction. And so, physiologists became convinced that voluntary movements are guided by these receptors, this sensory feedback. In fact this is a fallacy. If one tenses the muscles of the upper extremities, voluntarily, or electrically, one still would not have difficulty to reach any desired point in space. In fact, the movements are guided by vision when the eyelids are open and probably by joint receptors when they are closed. Yet, if one repeats the experiment by touching a position in space, reaching it by all kinds of approaches, one must realize the tremendous complexity of this act. Each combination of joint angles which is realized during the reaching of one particular point in space requires different patterns of activation of many diverse muscles. Moreover, even if one repeats the same general way of reaching this point, but changes the speed of movement, the distribution of nerve impulses to the major muscles must be organized differently each time. If in addition one takes a heavy object in one's hand, and still reaches for the same point in space in the same pathway, one must again change the succession of muscle patterns because of the changed inertia of the upper extremity. All these changes must be completed by the brain in milliseconds. Dr. Karl Smith of Madison, Wisconsin showed, by ingenious techniques, involving closed television circuits, that a delay of a few milliseconds in visual perception of the movement of one's hand completely disorganizes the handwriting. His technique also permits one to carry out the following experiment: the subject has to trace a pattern, drawn on a sheet of paper, which is reproduced by the television camera of a closed circuit. The subject is not allowed to look at the paper and is guided by the image of the paper seen on the television screen. By changing the position of the TV camera, one changes the apparent position of the pattern in space, but the subject has to correct the direction of his movements so that his pen continues to trace the pattern. This allows one to test the limits of the neuromuscular adjustment and create a type of temporary experimental sensori-motor disorder. Obviously then, a considerable part of the sensori-motor coordination of the voluntary movement is played by cerebral processes continuously computing the position in space of the moving extremity. This computing involves not only the segments of the extremities, such as the fingers, but any object the subject holds in his hand. Thus, one has no difficulty to "perceive," so to speak, different objects by the tip of the pen held in one's hand.

All these processes of guiding movements are largely unknown and considerable research has to be done for their elucidation. It seems that the problem would become much simpler if space were represented in the brain, and if one had the ability, even with eyelids closed, to make any point of the body or any point attached to the body, be it a pencil, or bumper of a car, to coincide with the desired location in space. From our point of view, we should learn how to improve this ability of spatial identification and how to do it in patients with physical disabilities.

The importance of the mental attitude in execution of a movement was revealed a long time ago in my experiment with maximum speed of alternating movements. Thus, for example, a rhythmic finger movement is more rapid when the subject tries to do successive movements of flexion, rather than extension, or when he tried to propel the tongue forward, than retract it backward, or tries to close the eyelids at a maximal rate rather than open the eyelids at that same rate.

Thus mental processes of computing spatial patterns and proper sensory feedback are major factors in guiding voluntary movements. Why is this of interest to us? We already mentioned that one may train the patient's propensity to recognize different points in space in order to develop computing abilities. What about sensory feedback? This appears to me as one of the major problems of research in neuromuscular disabilities. A method was recently involved by which the patient could perceive speech by securing to his chest different vibrators transmitting coded messages. Such vibrators could be used for other purposes. Suppose that we would like the subject to "perceive" the joint angle of his artificial knee. The lack of such perception leads to distortion of the normal gait pattern. It is relatively simple to build a device which, through a variable resistor with the knee joint angle, will modulate the intensity of vibration emitted by a vibrator attached to the chest of a subject. Thus, the latter will have a cue which will indicate to him that the knee is properly extended and, therefore, he may not fear to lower his artificial leg. Would it be possible then to improve the training of the AK amputee? Research will give the answer to this question. This, of course, may be done for any other joint of the extremities. Another problem is the intensity of the pinch produced with the help of a functional splint, sometimes activated by an artificial muscle. Here again a pressure sensitive strain gauge device attached to the tip of the finger may monitor a vibrator stimulating the chest of the subject and cueing him as to the critical intensity of the pinch.

Artificial feedback may be built into a muscle stimulating device. A switch in the shoe of a hemiplegic patient is arranged in such a way that when the patient lifts the drop foot from the floor, the peroneal nerve and tibialis anticus muscle are stimulated automatically, replacing the nerve impulses which are lacking due to central nervous system disease. We experimented with electrical stimulation of the upper extremity using electrically excitable muscles in order to activate function hand splints.

My original idea was as follows. Patients with electrically excitable but paralyzed muscles have a built-in muscular energy which can be used for activating prosthetic or orthotic devices. In an article in the Archives of Physical Medicine and Rehabilitation, Heather and Smith claimed that the energy of the cardiac contraction perceived by a sensitive device on the surface of the chest is sufficient to replace an artificial muscle in order to activate a hydraulic hand in a quadriplegic patient. My contention is that by continued rhythmic electrical stimulation of a muscle, remaining electrically excitable, the same effect may be achieved with considerably greater efficiency.

The energy of these muscles may, however, be insufficient for ambulation of the patient.

One of the more recent developments, still in a research phase, is that of a wheelchair capable of ascending and descending stairs. This will, of course, be a major landmark in rehabilitation of paraplegics and quadriplegics.

So much for the new research having for its purpose the therapeutic application of electronic techniques. Let us consider now the diagnostic implications of the recent progress in electronics. The major development in recent years is the beginning of the exploration of sensory nerves, spinal cord reflexes and inhibition, by combined methods of electromyography, electroencephalography and electroneurography.

During past years a technique of determination of motor nerve conduction velocities has been evolved. This at times permits one to differentiate the involvement of the anterior horn of the spinal cord from that of the peripheral nerve. Now, several techniques have been proposed for the determination of conduction velocity in sensory fibers, the most recent one based on the use of a computer device. This instrument permits the recording of electrical potentials, of very low amplitudes, provided that they constitute a response to a rhythmically applied stimulus. For example, if I stimulate the right median nerve twice per second, I am able to detect from the scalp, or in the upper cervical region after 3 minutes of stimulation, a potential, evoked by these stimuli. Just recently I examined a patient for whom these techniques were particularly helpful. He complained of a shooting, disagreeable sensation propagating from the ventral region to his right thumb. All the chronaxies were normal; no fibrillations were observed. Stimulation of the digital sensory nerve of the right thumb did not give rise to any potential in his median nerve, as it does in normal individuals. This normal response was observed on the uninvolved side. Moreover a potential resulting from stimulation of the radial nerve, sufficient to activate the brachial plexus was picked up on the corresponding area of the hand when the normal extremity was stimulated. No potential was observed when the involved extremity was stimulated in a similar way. A root lesion of the 6th segment was diagnosed.

Electromyography has also been used extensively for the study of the pattern of movements. For instance, a recent book by J. V. Basmajian on "muscles alive," summarizes new developments in this field. Dr. Long studied the motor patterns of the finer movements of the fingers and found that the long extensors and flexors on the one hand, and the intrinsics on the other, may be used interchangeably for the same joint motion under different functional conditions of the movement.

In our own laboratory, gaits were studied by high speed photography, accelerography and electromyography. In patients with above-knee amputation, a characteristic extra wave was found on the accelerograms of the involved leg. It is possible that by calling the attention of the amputee to the amplitude of this wave, it will be possible to improve his training.

A revolution in the recording techniques has been accomplished during recent years. Thus, miniature electrode amplifier-transmitters may be attached to the skin overlying muscles under investigation and permit one to study movements without having wires connecting the subject to the amplifiers and recorders. These telemetering techniques will permit one to eliminate artifacts from the wire movements, moreover the patient is able to move about freely in the room. I had one of these devices attached to my gastrocnemius and another one to the quadriceps. Relatively inexpensive FM radio receivers could easily pick up the signals. Such devices may be used for giving the patient auditory feedback from the contraction of a muscle which he wishes to exercise.

These are some of the studies which I believe will advance the field of rehabilitation of neuromuscular disorders. However, as I mentioned before, there is a considerable field for research in PM&R which does not require such refined equipment. Some day one will have to promote cooperative studies in order to test different rehabilitation techniques. For instance, one should find out whether electric stimulation is effective in hemiplegic patients. The results of such cooperative studies will have to be submitted to a complex statistical analysis. The latter may require the use of computer techniques, and therefore participation of highly specialized electronic engineers and biostatisticians, therapists, physiologists, pharmacologists and other representatives of the medical and behavioral sciences will be needed.

Isometric Exercises

My interest in isometric exercises was awakened by Dr. Elkins, a past secretary of the Board of Physical Medicine. He asked me about the claim that two minutes of isometric exercises per day may be as good as one hour of classical isotonic ones. Being an obstinate scientist I reserved my judgment until I could submit the problem to experimental analysis. I was joined in this work by a young Ph.D. student, Max, and a French neurophysiologist, Dondet, who temporarily wished to spend a few weeks in my laboratory. My experimental design was quite simple. I chose the abductor digiti quinti, a small hand muscle rarely used in daily activities, so that any increase of its strength would be due only to the experimental trials.

Results were not as dramatic as claimed by German investigators and several exercises per day were superior to just one daily exercise. However, these exercises were definitely superior to the isotonic ones. We explained the results as a greater percentage of time devoted to muscle activity in isometric as compared to isotonic exercises. It was essential to limit the isotonic maximal exercises to only 2 seconds at a time, to be repeated every 15 seconds so that diastolic blood pressure would not be affected.

There was a sad postscript to this study. Both collaborators for this project died tragically some years later. Dondet was the victim of a crash of a rapid French train, leaving an American wife and children. Max died suddenly from a heart attack, after having inaugurated a prestigious exercise establishment in the center of New York intended to prevent heart attacks in his patients.

The above discussion of normal volitional motor action with particular attention to gait naturally led to my own and other investigators serious attempts to alleviate disturbances of these physiological functions resulting from peripheral and central nervous system pathology. The use of substitution and activation methods by means of fundamental physiological techniques at our disposal represent the culmination of our accumulated knowledge.

Therapeutic Isometric Exercises

Inasmuch as one isometric contraction for 5 seconds each day in the hands of other physiatrists showed no better results than classical exercises, I devised a more precise approach to the problem. By using a small muscle, the abductor of the little finger, with the hand properly immobilized and with the contractions recorded, three series of subjects were studied. The first consisted of one contraction per sec., for 5 sec. once daily; the second was one per sec. for 5 sec., 15 times per day; and the third using a classical system of resistive exercises, which had been introduced into physical medicine by a Boston orthopedic surgeon. The results that I obtained were unequivocal. The first series of isometric exercises were not sufficient to produce a notable increase in strength or endurance of a muscle. The second series, which were repeated 15 times per day, gave results which were similar to those obtained with the classical technique, but with a spectacularly shorter time of effort. According to Hettinger and Miller, who had introduced isometric exercises in the practice of sports medicine, five second isometric exercises per day were sufficient to significantly increase the strength of muscles. They believed that the prodigious effects were due to a sort of neurogenic activation of mysterious metabolic processes. I could hardly believe this hypothesis and found out to my satisfaction that a much more physiological explanation could be invoked. It had been known for a long time that muscle contracts with greater efficiency if it is maintained elongated, or at least in an optimal length, and that super stretching is related to a degradation of performance. This is true for both innervated and denervated muscle. Also for many decades it has been known that muscle contains several receptors of the proprioceptive type. First of all the spindle. The spindle is represented by a formation of specialized muscle within the muscle having in its equator ramifications of sensory elements, the origin of centripetal influxes which were sent by the muscle to the spinal cord. Some of these messages were related to the sudden change in the muscle length; some of them were related to a fixed change in muscle length. This intrafusal muscle tissue was also under the influence of fine motor fibers; the gamma fibers, which originated from the central ventral horn of the spinal cord, the gamma motor neuron. Gamma motor neurons are able to contract or relax the spindle. In the contracted state the spindle is more sensitive to stretching and therefore more influxes per second would flow in the centripetal sensory fibers toward the sensors; the muscle itself being under control of the alpha motor neuron which sends axons to the end plates of the extrafusal fibers and contracts them under the influence of stimulation by the alpha motor neurons. Now the reflex arc which is formed under the condition of reflex would be the stretch of the spindle responding by a flow of stimuli arriving at the spinal cord and having only one synapse with the alpha motor neurons which indeed would send a contract-making message as if they wished to make the muscle resist the stretch. This is the stretch reflex studied in great depth by Sherrington. The stretch reflex involves motor neurons which otherwise cannot be mobilized by the effort of will of the subject. Sherrington called this situation the presence of a subliminal fringe. The motor neurons whose polarization was so great that whatever depolarization was brought about by centripetal messages from the cortex was insufficient to mobilize them into action. However, if partially depolarized by the stretch reflex, the voluntary effort could be sufficient to activate them. In other words, it was easily conceivable that without stretch, no matter how much our will to stimulate all motor neurons related to one particular muscle, we would fail unless the muscle was stretched, and unless the subliminal fringe was brought to bear on the final contraction. Indeed, we did ask the subject to contract his biceps with all possible effort when we recorded the electrical activity. When he contracted the muscle we asked him to contract the same muscle against resistance. In other words, when bringing the muscle to stretch, the amount of electricity generated was doubled, or even tripled by this maneuver. We felt this was the explanation of why resistive exercises, or isometric exercises were so effective, simply because in these cases we exercise practically all the fibers of the muscle. This would further explain why isometric exercises are superior to isotonic exercises. Moreover, our results showed that our exercises were also superior, or at least equal to the resistive exercises but with a considerable gain in time. Indeed, an exercise with motion, say flexion and extension of the biceps would be related to a length of the biceps of different values. When the biceps is shortened then obviously it works under conditions where the efficiency of the muscle is decreased. There is only one region of the range of motion in which the biceps would be optimally predisposed for mechanical advantage of all its fibers. Isometric exercise has to be done with the muscle placed in this area of motion without spending time to get into and out of this area. We felt that our explanation was logical, simple and conformed to the notion of modern neurophysiology and we presented it in several publications. It is only regrettable that many colleagues of ours have a limited notion of the problems involved as isometric exercises were accused of placing an increased load upon the circulatory system. Isometric exercises should be used for individual groups of muscles and not for all the musculature. Isometric exercises such as we prescribed should be used only for a limited time, only for 5 or 6 seconds, the time during which the arrest of circulation would be operational, and would not interfere with general cardiac action. Yet in order to prove that isometric exercises are inferior to some other exercises, the experimenter used long periods of contraction, and the involvement of large masses of muscle. We are convinced that when a physiatrist knows exactly what he is doing, the groups of muscles which are most needed at a certain period of time during the disease, such as the quadriceps and gluteus maxcimus, for locomotion, or the deltoid and latissimus dorsea for crutch walking will benefit from isometric exercise. If this limited objective were present in the mind of the prescribing physician our patients would only benefit from such prescriptions.

Another objection to isometric exercises was formulated; they may increase strength but not tenacity of the muscle. In the research which we did with objective measuring of the changes of strength of the muscle, we showed unequivocally that the tenacity statement was not true. The tenacity would definitely increase under action of isometric exercises. Obviously, isometric exercises are not favorable if the purpose is to increase the range of motion of a joint. Isometric exercises should be used judiciously, with a specific purpose in mind.

CHAPTER 8

Microsurgery

"He dreamed he was a physician healer
Of human sufferings, perceptive feeler;
Of drugs, a skillful magician,
Of surgery, a super-technician"

IN 1984 ONE OF the two hospitals where I practiced in New York (Cumberland Hospital) closed. At approximately the same time I received a note from a Dr. Terzis in Norfolk, Virginia. She learned that I was the American Editor of the International EMG Journal and asked me whether I could help her in locating an electromyographer to help her in her practice. I soon became aware that she was a reconstruction microsurgeon of some distinction. I read about this new, exciting type of work and decided to learn more about it. I called her, explained that I was only working half-time and might be interested in exploring the situation in Norfolk, and might even be interested in joining her myself, since Norfolk was only one hour by air from New York. Dr. Terzis gracefully agreed to see me in Norfolk.

What I found was most encouraging. Dr. Terzis, still a young surgeon with infectious enthusiasm, along with academic training and unbound energy was already writing her third book. She had succeeded in organizing in a relatively small medical center, a research laboratory and medical practice of great promise. Her Research Center of Reconstructive Microsurgery was located in the Eastern Virginia Medical School. She carefully planned and provided modern electrophysiological equipment, a teaching anatomical and microsurgical animal laboratory. She attracted a young well qualified neuroscientist who directed research of her clinical fellows, aided by a number of technicians. For example, in a project to which I later contributed as an electromyographer consultant, Dr. Terzis and her associates discovered, working on rats, that under certain conditions the introduction of a sectioned root into the lateral aspect of the spinal cord was followed by direct invasion of the root by ventral horn fibers, later making contact with the corresponding muscles. This experiment was independently confirmed by a group of Scandinavian investigators. In a more recent study of the same problem, to which I contributed my methodology of transcranial stimulation, more was learned about the future of this kind of surgery which aided both spinal cord and brachial plexus injured patients. In another collaborative study an ingenious following of fibers originating from various roots to the periphery on cadavers permitted them to map out important anatomical relationships that were helpful for surgery. Both studies attracted the attention of scientists, particularly in Europe.

I also visited her private office in downtown Norfolk. I was surprised to find outside of an academic hospital, a tremendously detailed workup of patients not only by a thorough clinical examination and associated EMG and radiological studies, but also by detailed behavioral testing and video tape documentation. I was surprised to see a score of patients coming from distant states, as well as from Europe, Asia, Mexico and the Middle East. During my long EMG practice I had not seen as many brachial plexus victims as I did during that visit. The patients spent long hours there going through all these tests to be finally examined by Dr. Terzis. I was impressed by the infinite trust that she could instill in her patients, seeking her help as a last resort. I was also surprised by the considerable work capacity of the doctors who worked in her office from early in the morning until 11:00 p.m. She was helped by an administrative assistant and physicians devoted to her. Sometime later she said she never fired a secretary, but occasionally they did leave her. I found that some of her surgeries lasted from early morning to late at night, sometimes for 24 hours. She never missed her work the next morning. I was therefore won over by Dr. Terzis and progressively shifted my practice from New York to Norfolk. There I developed new methods of diagnosis and therapy for her patients. So I remained there for six years, actively involved in the program of Dr. Terzis. I learned a great deal. You may have had an experience as a child trying to catch a lizard by its tail only to find out that this little green reptile, by a mechanism that remained incomprehensible to us, separated itself from its tail by an action of its will, and its physiological activity, and escaping, wiggling down some yard in front of us. It left its tail in our hands. I might have become sorry for its future tail-less existence but was told that the lizard had an innate ability to regenerate its tail. Unfortunately, we mammals lost this ability to regenerate an organ during our evolutionary struggle for existence. And yet, our nerves and muscles, as well as other tissues, do have the ability to regenerate themselves. This involves a marvelous mechanism not yet fully understood. Whenever a mammalian nerve fiber is injured, by severance (knife, or bullet) the corresponding nerve cells of the spinal cord are alerted as they no longer receive data in a normal way. Distorted signals do arrive at the motoneurone and inaugurate a response. Microscopic examination shows that the cell becomes the scene of some precipitated activity resulting in the appearance of fragments of material, called Nissl bodies; shortly thereafter the proximal (central) fragments of the injured nerve starts emitting sprouts that actively search to penetrate into the distal segment of the injured nerve. It is well known that each actively conducting nerve fiber is surrounded by a myelin sheath which is part of the Schwann cell; these latter cells are also alerted by the motoneuron and repair of the disrupted myelin sheath is inaugurated.

However, there is a problem. Each nerve is made of a large number of fascicles, some of them being motor in origin, some of them sensory (originating in the bipolar cells located in the dorsal root ganglia). Obviously if a motor axon regenerates and pushes its new sprout into an empty tube it may be "misguided." One has to keep in mind that indeed after a complete transverse section of the axis, the segment located in the distal part perishes by malnutrition because it receives no nourishment from either the motoneuron in the spinal cord, or from the spinal root ganglion. This is the Wallerian degeneration. However, the "tube" in which the axon succumbs remains. If a degenerated motoneuron sprout penetrates the tube of a sensory fiber it will happily grow in this tube because it now receives nourishment. This happy mating persists until it reaches the terminal peripheral receiving element; the receptor element sends signals to the motoneuron that it can not interpret, and the motoneuron signals sent down will be ignored. Therefore this unit is functionally non-operational. However, fortunately the motor axon that does survive in a motor tube is able to increase the number of its terminals and the activated muscle is capable of hypertrophy by a factor of five times.

More often than not a sprout arriving from the central segment of the injured nerve will encounter scar tissue instead of a warm bed of the remaining axon tubes. A very irrational and complex growth of the central segments then takes place with sprouts intertwining with fibroblasts and collagen and giving rise to a neuroma, which is often the source of intense pain. Wasteful regeneration results from this occurrence and a tumor may form at the site. At times it may be necessary to remove the tumor and suture the two nerve ends correctly.

However, in some injuries the trauma is so violent that the two nerve segments become widely separated by a distance too long to be able to be sutured effectively. In the past, in some instances, the two nerve segments were forcefully joined. Millesi and Terzis showed that this forced "sad gown" marriage of the two endings did not work well. The modern microsurgeon interposes another nerve from the same patient between the central and peripheral segments. In most cases a segment of the sural nerve is utilized. The sural is a peripheral sensory nerve which contributes little to the function of the lower extremity. Several strands of the sural nerve may be used, as the number of the tubes offered by a single strand may be deficient for that required by the injured nerve. Sometimes one uses the saphenous and other sensory nerves without a major functional loss. In cases where the eighth cervical and first thoracic are hopelessly damaged one may use an entire ulnar nerve to make such grafts. At times one may use non-vascularized and sometimes vascularized grafts. According to Dr. Terzis, who popularized these procedures, and other investigators, the vascularized grafts permit a more rapid transport of the regenerating nerve, particularly if the "bed" in which the injured nerve lies is poorly vascularized to begin with. Millesi indicated that the graft is more successful when the gap between the two segments of the injured nerve are relatively small. The situation is less favorable when the graft is 5 to 10 cm or longer. However, I found that in some of Dr. Terzis' patients, in whom the final regeneration was functional after several years, that recovery did occur despite very long gaps.

Success also depends on the time elapsed from the injury prior to surgery. Also the age of the patient is extremely important. In infants and children the grafts are particularly successful for two reasons; the relative shortened distance that regenerating axons need to travel and the greater regenerative capacity of children's nervous tissue. A greater plasticity of the brain allowing use of the grafted nerve, as soon as possible, may be a third factor for the more efficient regeneration and superior outcome of grafted nerves in the young.

Two particular domains of traumatic peripheral neuropathy requiring microsurgery are: 1) brachial plexus palsy; and 2) facial nerve palsy.

Brachial Plexus

The spinal cord communicates with the muscles and skin of the shoulder, arm, forearm and hand via the roots emerging at the level of C-5, C-6, C-7 and T-1. Since there are only seven cervical vertebrae the C-8 root emerges below the C-7 vertebra, and the T-1 root below the T-1 vertebra, etc.

The T-1 vertebra may be identified by the fact that it is connected with the first rib. The roots C-5- C-6, C-7, C-8 and T-1 interlace with one another to form the brachial plexus. Anatomists found an order in the plexus and established the following "landmarks": The 5th and 6th cervical roots unite to form the upper trunk. The C-7 cervical root is continued by itself as the middle trunk. The C-8 and T-1 roots combine to constitute the lower trunk. Each of the trunks give off a twig or division; these divisions form the posterior cord. The main portion of the upper trunk becomes the lateral cord. The medial (internal) root of the median nerve derives from the medial cord. The lateral root of the median nerve derives from the lateral cord.

A twig from the upper trunk innervates the supraspinatus muscles and the infraspinatus muscle via the suprascapular nerve. A branch from the posterior cord forms the thoracodorsal nerve that innervates the latissmus dorsi muscle. Finally the lateral cord gives rise to the musculocutaneous nerve that innervates the biceps and brachialis muscles, the pride of male muscular beauty. The pectoralis major, another proud muscle of the male, is innervated by small branches of the plexus arising from different roots. In women these muscles are concealed by the breasts.

It is quite fascinating to become cognizant of the destination of all these nerve branches. The hand receives supplies mostly from the C-8 and T-1 roots. The shoulder is innervated predominantly by the upper trunk, deriving from C-5 and C-6 roots. The biceps is "fed" mostly by C-6. The extensors of the wrist and fingers are fed mostly by the C-7 root, except the extensor carpi radialis which is supplied by the C-6 root. The extensors of the terminal phalanges of the fingers are supplied by the ulnar nerves.

Changes in the Internal Structure of the Constituents of the Brachial Plexus

At the root level in most cases one finds one big fascicle occupying the entire root. Then at the trunk and particularly the cord level the fascicle is broken down into many smaller ones. Within each one of the fascicles the nerve fibers follow a spiral path. For example, the nerve fibers destined for the suprascapular nerve located at the posterior side of the fascicle of the fifth root; but then these fibers push away from the posterior side and try to be located externally to the point where they leave the upper trunk to innervate the muscles. Why? Because on that level they have to detach themselves from the external aspect of the upper trunk toward the supraspinatus. One can ask oneself, again, "Why?" Why could not these fibers be located at the beginning at the lateral side of the fascicle? One does not have the answer to this question, except obviously when one chooses the spiral path, one becomes longer. For what reason nature decided that it is better to be longer than shorter is a moot question. So each nerve getting out of the plexus to join the terminal nerve dances this intrafasicular waltz for an unknown reason. Probably the final intraplexus organization reflects developmental events relating to the formation and migration of the upper limb myotomes and the supplying peripheral nerves.

Let us now review in some detail the function of the muscles that these different formations of the brachial plexus innervate. The main muscles of the shoulder are: adductors, abductors, flexors, extensors, elevators and external rotators; that is bringing the arm to the side, in front, behind and above the body. Abduction is accomplished by a single muscle, the deltoid, which is aided by the supraspinatus. The adduction (bringing the arm to the body) is accomplished primarily by the pectoralis. The arm rotation (external and internal) is accomplished by the small muscles such as the infraspinatus, teres minor and teres major; the serratus also support other important functions of the shoulder by assisting in the final degrees of shoulder elevation. One should not forget that the upper extremity has no bone connection with the body except through the relatively tiny clavicle. The humerus, the bone of the arm, articulates with the scapula, articulating itself with the clavicle, which in turn articulates with the sternum. The fact that the upper extremity appears to be firmly connected to the body results from the powerful muscles and ligaments that hold it against the thorax and to the other muscles that suspend it from above.

The main muscles of the elbow are the biceps (the flexor and supinator of the forearm) and the triceps (the extensor). Both of these muscles are rather "stupid" with each one having only one simple function, to bend the elbow in and out, with the exception that the biceps has a supination action. Supination consists of rotation of the hand with the palm up, the beggar's hand position. When the biceps acts like a supinator it develops the maximum of its activity. A relatively small muscle, the brachioradialis, innervated by the radial nerve, also helps to flex the elbow when the hand is half pronated and half supinated. The triceps moderates the action of the biceps if it becomes too violent; it prevents one from beating his own chest. It also serves to hit an adversary with an outstretched arm.

As I said before, all of these muscles are innervated by the upper trunk, derived from C-5 and C-6 with some participation of C-7 for triceps and latissmus dorsi.

The extensor muscle of the forearm, the extensors of the wrist and of the first phalanges of the fingers are all supplied by the radial nerve, which is derived from the C-7 root, from the C-6 root (extensor carpi radialis) as well as from the C-8 root (extensor of the thumb). Extensors of the wrist are a luxury for the patient dramatically reduced in functional capacity. One can fuse the wrist and use the surviving muscles for moving the fingers by reconstructive surgery.

The flexor mass consists of pronators (the uppermost in its innervation), flexors of the wrist, flexor carpi ulnaris flexor carpi radialis, the first innervated by the ulnar nerve, and the second by the median nerve. The flexor sublimis flexes the middle phalanges of the fingers, while the flexor profundus the terminal phalanges. Here again the flexors of the wrist are "luxury" muscles and in the patient with reduced function and a surgically fixed wrist, these muscles may be used for other purposes, such as helping to flex or extend the fingers. The flexor mass is supplied by the C-6, C-7 and C-8 roots via the median nerve while the flexor carpi ulnaris and the ulnar part of the flexor profundis that are innervated by the ulnar nerve.

Finally the "intrinsic" muscles of the hand have ultimate control of the thumb's palmar abduction and opposition (supplied by the median nerve); of thumb (key pinch); of flexion of the proximal phalanges of the digits and extension of the distal phalanges (ulnar nerve). Among these different movements performed by the intrinsics, each of them perishing because of the lack of regeneration of the nerve fibers so far from their nursing mother, one can extend and flex the fingers by transferring the tendons of the surviving muscles of the forearm. The patient would not be able to play the piano but would be able to assist the normal hand in grasping and holding objects.

We shall now describe in the same detail sensory innervation. Of course the sensory fibers are useful for the shoulder and for the forearm, particularly when they are carrying information from the muscles, however there are many fibers innervating the upper extremities originating from the thoracic spine, and this deficiency is therefore not vital. It is important, however to realize that the palmar skin of the hand derives its sensitivity from two nerves: mostly from the median, supplying the skin of the thumb, index finger and one half of the ring finger, the other half and little finger being supplied by the ulnar nerve (which is vitally important). The radial sensory nerve may also be discarded in these handicapped individuals. Now that we know the function and value of the different roots and nerves, we may consider various injuries of the brachial plexus. Two major types must be considered: 1) obstetrical palsy of the infant; and 2) traumatic lesions of the adult.

Obstetrical Palsies of Infants

These were described some two hundred years ago. Three main categories are established: 1) Erb's palsy, involving mostly the upper roots; 2) Klumpke's palsy, expressed by the lesion of the C-8 and T-1 roots; and 3) Global paralysis, which refers to lesions of 1) and 2). In fact, my encounter with infantile obstetrical palsy in combination with Dr. Terzis revealed that in a substantial number of instances the C-7 root was involved. Parenthetically one must mention that Dr. Klumpke was one of the first women physicians in France. The above division is important as in Klumpke's palsy the hand is primarily involved (both paralyzed and insensitized), while in Erb's palsy the hand functions might be preserved.

It is important to recognize that the major difference between obstetrical and adult palsies is related to the growth of the extremity. In adults the involved upper extremity retains the same dimensions because the palsy occurs after the termination of skeletal growth, while in infant palsy one of the dreaded consequences is the presence of arrested growth so that in a child and adult with a history of obstetrical palsy the hand cannot be hidden in a pocket but remains grotesquely small, sometimes only two thirds that of the normal. This of course causes considerable psychological trauma in children, in addition to the physical disability. On the other hand, in the obstetrical palsies if treated early, the upper extremity can be reconstructed more successfully because of the superior regenerating capacity of the infant and child, and because of the greater plasticity of the young brain.

One can imagine the grief of young parents, when after a usually normal pregnancy, they discover their eagerly awaited child cannot move his or her arm. Often the child is unusually plump and the labor has been difficult, sometimes with the application of forceps, but not necessarily so. In most instances the infant's disability comes as a complete surprise. Obviously the first reaction often is to accuse the obstetrician; but in fact in many cases no obvious obstetrician fault can be discovered. Then there is a period of self accusation during which they assume an imaged attitude of guilt. Fortunately, about 4 out of 5 of the cases of obstetrical brachial plexus palsy improve spontaneously. My former resident, Dr. Zalis, described many of these cases where an EMG alerted the physician as to the probable future improvement, the muscles being inactive because the initial injury left the child with a lack of habit of voluntary muscle control. It is therefore important to conduct an EMG as early as possible. In some of these infants, electrical stimulation with brief stimuli carried out with conditioning training, may allow the child to recover without surgery. However, 20% of infants do not recover spontaneously; this forces the physician to agonize whether to operate on a child who may make further progress or decide to wait until the child matures. The general opinion at the present time is that in cases of Erb's palsy lesions, an experienced microsurgeon should operate at three months of age if there is no functional activity of the biceps. This muscle appears to be crucial for the physician's decision as will be clear from the following notes. However, prior to surgery one must obtain a detailed EMG examination, despite the fact that infants more that adults are extremely disturbed by this testing, since the needles prick and the electric current bites.

It is now time to discuss the essentials of electromyography. The most important notion of electromyography concerns the definition of the motor unit. This is a motor neuron sitting in the ventral (anterior) horn of the spinal cord and its motor appendix (an axon) that may be longer than one meter in length, as well as its terminal twigs which innervate each muscle fiber. There may be thousands of these terminal axonal twigs in a muscle such as the quadriceps, or the gastrocnemius, or even the biceps brachialis, or only ten or twenty in the small intraocular muscles. The facial muscles form an intermediate picture. The more a muscle is involved in a fine selective movement, the less axonal terminal ramifications are observed. Now imagine an EMG electrode in the midst of all of these fibers innervated by a single axon. The needle electrode picks up electric potentials surrounding it. Even if the patient reduces his effort to a minimum and succeeds to mobilize voluntarily only one unit, it requires a finite time (conduction time) for each muscle fiber potential to reach the tip of the needle which is insulated except at the tip. A muscle fiber conducts electricity with a velocity of four meters/sec, or four millimeters in one millisecond (msec). If one of the fibers belonging to the same motor unit is two cm distant from the tip of the needle, its potential will be recorded five msec after the time of initiation. Granted, only axonal ramifications receive their command almost simultaneously. But the recorded motor unit potential is a sum of all the muscle fiber potentials supplied by one axon and therefore the arrival times to the tip of the needle are dispersed over at least five or more msec. Therefore, a normal motor unit potential is in fact a statistical expression of the distance separating the needle electrode from different (thousands) of muscle fibers surrounding it. If this axon is diseased some of the terminals will not conduct. Some of the sprouts will innervate muscle fibers that normally do not belong to this motor unit. As a result the dispersion in time of the recorded potential will increase, so that the regulation of the motor unit potential may become, for example, fifteen msec instead of only five. In addition, some of the territory of the motor unit will be covered by the dead fibers so instead of a normal statistical distribution of the time of arrival of muscle fiber potentials to the needle there will be some missing portions of this distribution and therefore the motor unit will become polyphasic. Finally, some of the axons, if not all, become nonfunctional and the corresponding muscle fibers remain without any voluntary or central control. They starve from the absence of the protoplasm flow that usually reach them from the spinal cord. As a rule the membrane becomes irritable and unstable and the single muscle fiber pulsates independently. These pulsations are expressed by fibrillations or by positive spikes.

Moreover, if the myelin sheath degenerates the conduction velocity along the fiber changes, one finds stretches on nonmyelinated fibers, a process called *segmental demyelinization*. As we saw before, normally the electrical potential transmits excitation from one part of a nerve to another part by jumping from one node of Ranvier to the next. In this way the conduction velocity ranges around 60 to 70 meters per sec. However, in the nonmyelinated fibers the velocity is only 2 meters per sec. When a jumping potential finds a stretch of nonmyelinated segments in the sick axon it can not jump to the next node of Ranvier that is located too far away. Instead it has to slowly swim with a velocity of only about two meters per sec. Therefore in a pathological nerve fiber with segmental demyelinization the excitation is transmitted by at times jumping, and

Microsurgery

at times swimming with an average reduced conduction velocity. If the conduction velocity is found to be 30 meters/sec, it indicates that about half of the nerve fiber is demyelinated.

Now with these essential notions the electromyographer first determines whether or not the particular muscle exhibits any voluntary potential. This examination should be accomplished as soon as possible post-injury. The motor unit potentials usually "interfere" or inter-relate with one another under normal conditions; the resultant pattern is called an interference pattern. An incomplete interference pattern is frequently observed in the presence of even relatively minor nerve injuries. A good voluntary activity is an index of a limited injury. When one finds only isolated motor unit potentials one must conclude that there is a severe injury of at least some nerve fibers. When no motor unit potentials are recorded one concludes that a transection of the nerve has occurred, or that there is a functional transmission block. Under these latter conditions the electromyographer usually can identify abnormal potentials such as polyphasic waves or fibrillations. In the case of a blocked nerve the nerve continues to transmit electrically induced potentials. If however, after five days there are no voluntary potentials and there is no appearance of fibrillations below the putative lesion, one must conclude that the lesion is total and that the nerve is transected. If one waits another fifteen days the appearance of fibrillation potentials and positive sharp waves confirms the diagnosis. It is obvious that these fibrillations, as important as they are, still do not bring any "new" information to the electromyographer.

In the past, traumatic lesions of the nerve have been classified as follows: (1) the presence of a block; (2) the presence of axonal involvement; (3) total injury to the entire nerve. This of course is a helpful concept provided that one realizes that the nerve is not a single unit, but is made of thousands of axons. In the same injury some of these axons may be transected, others may only be blocked; and some others may be spared. Because of this fact the electromyographer may find in the same muscle, after injury: (1) almost intact interference pattern; (2) practically normal fast conduction velocity; and (3) fibrillations.

When an expert gives a report of this type to the surgeon, the latter usually suspects the competence of the electromyographer, and believes that he does not know what he is talking about. In fact, unfortunately, this is the other way around. The best attitude is to choose an experienced electromyographer, and then to trust him or her when they draw conclusions from the examination.

When the nerve regenerates the situation becomes somewhat more bewildering. The best indices are: (1) reappearance of motor unit potentials; (2) moderate voluntary activity; and (3) the appearance of an interference pattern.

Because the nerve and the muscle are not single units, one may observe simultaneously very sick potentials such as polyphasic and fibrillations, at the same time as normal ones. The most important thing is the reappearance of motor unit potentials, as well as later the appearance of complex voluntary potentials.

The conduction velocity represents another source of puzzlement for the inexperienced physician. For example, in a lesion in which a few fascicles have been spared, the maximum conduction velocity may be normal, but when most of the remaining nerve regenerates, the electromyographer records the appearance of a slower conduction velocity. This is an encouraging sign for an informed physician. This will become obvious later.

In examination of the upper extremity one may start from the hand and work toward the neck or from the shoulder down. One chooses either technique according to the degree of involvement of the hand. If the hand is almost normal (Erb's palsy) one starts the test from the shoulder. If the hand is deprived of sensation (Klumpke's palsy) as well as mobility, one starts at least temporarily from the hand. During this examination the electromyographer has to keep in mind the most important questions of interest to the microsurgeon, "Are there any able roots, or only damaged ones?" and "Which ones are avulsed, and which are still connected with the spinal cord?" Unfortunately, in the infant the electromyographer is unable to perform one very important test helpful in revealing root avulsion lesions: this is the lamina test, which requires active participation of the patient. In this test the patient must report sensation in the fingers following stimulation at the level of the existing roots. However, this deficiency can be partially overcome by pinprick studies during sleep in which the movement of the extremities, as the result of stimulation of the distal parts, can be reliably quantified.

There is, however, a good test to assist in the diagnosis of avulsion of either the C-8 root, or T-1 root, or both; C-7 root avulsions are very difficult to differentiate from the C-6 root avulsions. This test consists of stimulating the ulnar or median nerve at the wrist and recording sensory potentials from the fingers or the hand. Occasionally one may also stimulate the radial nerve. However, this test has many restrictions as the reader may readily understand. First of all it is only applied to those in-

Microsurgery

fants in whom one or several fingers have lost their sensibility, and in whom the intrinsics are paralyzed, affecting either the ulnar or median nerve, or both. The test is positive when stimulation of either the ulnar or median nerve elicits normal sensory potentials derived from the fingers. The potential derived from the fifth finger is concured by the ulnar nerve and therefore is of interest in the suspected avulsions of either C-8 or T-1 root, or both; the sensation of the thumb is controlled by the integrity of the C-6 or the C-7 roots. The sensation of the midfinger depends upon C-6 or C-7 roots also. Now, why is this test so important?

You must remember that before the roots, ventral and dorsal, form a spinal nerve by uniting one with the other, the dorsal root swells as it contains the spinal ganglion which harbors so-called bipolar cells; one pole corresponding to the twig that penetrates the posterior horn of the spinal cord and another which is the origin of the sensory axon, sometimes more than a meter long terminating in the skin of the corresponding fingers. The glabrous skin of the thumb is generally innervated by the C-6 root, the skin of the little finger by the C-8 and T-1 roots, and the skin of the intermediate fingers by the C-6, C-7 and C-8 roots. Now avulsion of the root resulting from a lesion proximal to the spinal ganglion causes the root to be avulsed together with the spinal ganglion. In such cases, providing that there is no additional lesion *distal* to the root, if there is any part of the brachial plexus and the peripheral nerve intact the avulsed root continues to "feed" the digital nerves several yards from the lesion and therefore the electrically induced potentials are of normal amplitude and latency following stimulation of the nerve at the wrist. In such cases an electromyographer may claim that there is indeed an avulsion of one or several neighboring roots. However, if the test is not positive an experienced electromyographer should never say that there is no avulsion of the corresponding root. Indeed, an avulsion of the root may coexist with an additional lesion of the structures distal to it and therefore interrupt the nourishing fibers from the digital nerves. In these cases the digital nerve potential will not exist, and yet the root may be avulsed.

Unfortunately, an inexperienced surgeon may claim the electromyographer is wrong if the surgeon finds an avulsed root despite the EMG test being negative. Fortunately, myelography and various CAT scans may contribute to the diagnosis of an avulsion by showing a protruding meningeal sac at specific levels of the spine; when the picture is clear, this is usually indicative of evulsion. However, one should keep in mind that myelography identifies the levels of the exits of the roots, not their function. The EMG identifies the function of the corresponding root. It is well known that in a certain proportion of normal individuals the spinal cord is either pre- or postfixed. In other words, the C-7 root may innervate the intrinsics and the C-5 may innervate the phrenic nerve. When there is a discrepancy between a clearly noted finding of the myelogram and the positive EMG test one should always consider anomalies.

In cases of negative result of this test one must place the needle in the cervical paraspinals. They derive their innervation from the roots immediately after the formation of the spinal nerve and therefore distal to the spinal ganglion. If in a young infant paraspinal muscle EMG shows fibrillations or positive waves it means that the lesion is proximal to the twig that innervates these muscles. If these muscles show potentials in the face of distal paralysis, it implies a more distal rupture. Unfortunately at this level one cannot localize with certainty, nevertheless the information may be of some use.

The EMG and certainly the muscle potentials from the paraspinal and the intrinsics takes into consideration the following:

1. The presence of potentials in a muscle must be derived from the corresponding root. If potentials are derived during the examination, of the usually struggling child, one must grade the results as follows:

    a. No potentials. The corresponding root is either severely damaged, or avulsed.

    b. Only motor unit potentials are seen in isolation. Serious injury but some continuity of the nerve fibers.

    c. A moderate amount of activity.

    d. A high amplitude activity but not complete interference.

    e. A complete interference pattern of the top of the line.

In principle, if one can be sure that involved activity is present, even only as motor unit potentials, one should state that there is no complete avulsion or rupture of this particular root, or that this particular root is not completely damaged. However, from the surgical viewpoint this means that this particular segment of the brachial plexus cannot be made functional. It should be remembered that anyone of the remaining axons may give rise to isolated motor unit potentials. In our experience some such patterns would not prove to the surgeon that continuity exists between the spinal cord and the muscle, but one should not be puzzled by such discrepancies after exploratory surgery.

The most important example is the EMG of the supraspinatus. This EMG shows either new motor unit potentials, or little "voluntary" activity. The authenticity of this finding is incontestable. Indeed, potentials originated in the trapezius may invade by volume conduction the EMG of the supraspinatus, but then a single unit potential would not be observed. However, if the potentials are quite impressive and the clinical evidence of the presence of the supraspinatus function (abduction) is absent, one may make an additional "checking." Indeed, if the stimulation of Erb's point shows an abnormal latency, then these are definitely not potentials derived from the trapezius. One can also stimulate the spinal accessory nerve and find out whether the needle reveals evoked potentials from the supraspinatus. If it does, the results are also incontestable. One should understand the importance of the supraspinatus test because this is a muscle which is indisputably supplied by the 5th root and if there is a good EMG from the supraspinatus, this means that this 5th root could be further utilized by the surgeon for grafting.

Next one explores the deltoid muscle which may receive its innervation from three different roots; C-5, C-6 and C-7. Again the EMG of this muscle is essential, as a weak deltoid cannot abduct the extremity. Therefore the clinician tends to disregard a possible connection of this muscle with the spinal cord if the abduction can not be obtained clinically. The conduction velocity is of course of considerable significance in this case. When the evoked potential is present the connection with the spinal cord is incontestable. The degree of flowing will indicate the extent of the injury to the nerve fibers.

The next muscle to explore is the serratus. One of the muscles innervated by several roots, but the twigs are derived from the roots very close to the spinal cord. Unfortunately sometimes it is difficult to identify these potentials.

The next muscle to study is the latissimus dorsi. It is a most remarkable muscle. It is initiated in the brachial plexus, but its lower insertion is at the iliac crest of the pelvis. It is therefore one of the muscles which in the future may be used for the reconstruction of the lower extremities in paraplegics. In patients with brachial plexus palsy it is often used for reconstruction of the biceps or triceps as a pedicle transfer, or as a free transfer. Its function is to elevate the body to a sitting position with an extended arm. It is not essential for the handicapped individual to have both muscles present on each side. The repair of the biceps is by far more important.

Finally, one places the EMG needle in one of the most important muscles of the upper extremity, the biceps. This muscle functions as the hand positioner at the variable levels in space, and it flexes the forearm. Again, it is important to discover the degrees of innervation and the conduction velocity. If there is no innervation the possibility of C-6 avulsion should be considered; in many cases the root is only damaged. Of course, the combination of a normal sensory potential at the thumb with a paralyzed and insensible hand would favor the diagnosis of avulsion. One should be aware that the presence of a strong triceps may stimulate the voluntary potentials of the biceps, since each time one functionally contracts the biceps one automatically contracts the triceps. For an infant with any damage of C-6 a contraction of the biceps is a real effort; therefore one should be sure that the triceps is not the origin of the potentials observed in the biceps. The situation is complicated by the fact that when one tests the conduction velocity of the musculocutaneous nerve one must stimulate the point which is responsible also for the contraction of the triceps.

The "extensor mass" in most infants is paralyzed because the region of C-7 is frequently involved in obstetrical palsies. The "flexor mass" depends on C-6 and C-7. This is an extra test. In case of a more or less functional hand, some of these muscles depend on C-8. At the end one reaches the examination of the intrinsics in order to determine whether the hand is spared. The microsurgeon evaluates all the above electro-clinical data together with clinical video demonstrations.

CHAPTER 9

Functional Electrical Stimulation

"Enough: With a renewed mind
He'll set out a novel course
He'll try now to seek and find
A new purpose to endorse"

OF ALL THINGS in my scientific past, functional electrical stimulation stands out so far as my most fruitful endeavor. I know that in order to achieve what I have done in this area, I stood on the shoulders of giants. Recently I attended a meeting in Vienna on functional electrical stimulation and was fascinated to witness the full realization of my work. I was very grateful to the many workers in this field who carried out what I had left undone.

In order to understand the work that I did in this area, one should go back to the time when I was a student in the laboratory of Wedensky, the great Russian physiologist of the last century. He demonstrated more than one hundred years ago, his "electromyophone" that he built using the then recently introduced telephone. The present generation has difficulty in accepting that such an achievement was possible more than a century ago. And yet it is a matter of record that Helmholtz recorded conduction velocity in the human median nerve as precisely as we do now. He even included a warning to take into consideration the temperature of the skin! Just think of it. Some obscure finding of our contemporaries may serve as a basic notion of prodigious future developments, one hundred and fifty years ahead.

During studies at the Salpetriere, I revealed the rhythmogenic excitability in human nerves and proved, contrary to the then classical doctrine, that a human nerve may respond continuously and rhythmically to the passage of direct galvanic current. "Give me a current of more than 2 micro amps," I said, "and I will make this nerve fire repeatedly." A notion that was confirmed later by others.

I also came to Paris with the knowledge that Nicholas Bernstein, a neurophysiologist and mathematician from Moscow, had developed a sophisticated physiographic method of studying gait. The instantaneous values of accelerations of different body segments permitted Bernstein to understand, more than anyone before him, the biomechanics of gait. Unfortunately for Bernstein, the electromyographic technique was not yet available.

In Paris I wondered whether new technology could provide us with the instantaneous values of accelerations of different centers of gravity of body segments which would then save us days and weeks of labor. I succeeded in developing a piezo-electric accelerometer for this purpose based on the discovery of the physicist husband of the celebrated Madame Irene Curie.

Later I developed my methodology of the study of gait based on these technical achievements using the Bernstein electric lamps. I used three gauges which are much easier to deal with, and also many channels of electromyographs and recordings of the rotations of main joints. I introduced the notion that gait is based on the presence of an invisible biological field. Many investigators deny that biology could have invented the wheel, but in fact it did. The wheel mechanism may be considered in conjunction with the interplay of the forces of gravity of inertia of moving muscles, of friction and of reaction from the road. Vehicles use wheels as a means of displacement and it is useful to consider human locomotion in comparison to a wheel. The spokes of a wheel can be represented as a needed leg and the segments of the wheel can be divided into foot-like plates. If all but two of the spokes are eliminated, the function of the wheel can still be performed, but provided that while one spoke foot component is on the floor, the other is transferred forward from behind, so that the next step can be taken. Thus the succession of the stance may be compared to an animated biological wheel. I then learned that by a little cube of tissue on each anterior horn of the spinal cord, a symphony was played by the nuclei of the soleus illiacus, gluteus maximus, quadriceps, hamstrings, gastrocnemeus and tibialis anticus, to name only the major players, aided by the abdominal muscles and the rectus spinalis.

I also found, on the basis of long experience, that brief pulses of electric current,, the bigger the better, could activate all of these muscles, if only the subject was able to mobilize them in the proper order.

Functional Electrical Stimulation

Fortunately, at the end of the 50's, the breakthrough in electronics occurred with the invention of transistors. This considerably simplified my task inasmuch as a portable stimulator could be built. However, such portable instruments at that time were not available and I could not use them for my studies. Therefore, with the help of Dr. Franklin Offner, a manufacturer of EEG equipment, I was able to demonstrate my idea on a limited scale. I took for a model, the foot drop of a hemiplegic patient. I placed a switch in the shoe. Each time the patient would lift his leg from the floor, a current would be initiated and would stimulate the tibialis anticus and the peroneal muscles. The tibialis anticus elicits dorsiflexion of the foot and its inversions. The peroneal muscles produce dorsiflexions and eversions, infraeversion and eversion cancel each other and the whole movement is expressed by strict dorsiflexion. Thus a closed loop was created and an automatic correction of the hemiplegic gait was achieved. Obviously the same switch or one placed in the other shoe could energize the quadriceps should the latter be weak.

I demonstrated my method in many conferences and hospitals. I would say, "Now the current is up and you see this clumsy effect of the foot drop impeding the ambulation of the patient. Now when I turn the switch on the patient walks with the gait of a soldier. Now I am going to turn the switch off and the patient will return instantaneously to the clumsy foot drop gait." On one or two occasions, the patient did not resume the clumsy gait, and the disappearance of the foot drop was seen for about ten to twenty minutes afterward. Following some meditation, I concluded rightly, that this remarkable after affect resulted from inhibition of the spastic gastrocnemius that outlasted stimulation. Functional electrical therapy, as I called it in my first paper, involved both activation of the agonist and inhibition of the antagonist. In a paper I delivered at the International Congress of Physical Medicine and Rehabilitation in 1960 I reported that functional electrotherapy may be conceived in order to permit the paraplegic patient to stand or perhaps walk and hemiplegic patients to correct their faulty patterns, apprehension and gait . . . so that at the very time of the stimulation, the muscle contraction has a functional purpose. This paper, which was published in the proceedings of the International Conference in Washington, was republished the next year in the Archives of Physical Medicine and Rehabilitation.

These then-revolutionary ideas proved to be correct, but it took thirty years for their, at least, partial realization.

Fortunately, functional electrical therapy, renamed functional electrical stimulation incited surgeons to try their craft to move this methodology forward. Unfortunately, in many cases the wires around the nerves slightly damaged the nerve tissue and the patient had to be reoperated. The system used by surgeons was to introduce inside the body the receivers which were activated from outside by induction and therefore the skin was not permanently penetrated. This tradition was later broken and the wires were directly introduced under the skin.

The major problem was the difficulty to stimulate deep-seated muscles such as the soleus and gluteus maximus. The soleus is very deeply seated and it cannot be stimulated from outside. On the contrary, the gluteus maximus, an extraordinary versatile muscle with extreme power, can be stimulated, but in many cases it is painful. I welcomed the efforts of surgeons but no surgical solution was forthcoming. And so later on, I developed a simple electromechanical brace which simulated the action of both the soleus and gluteus maximus. Relatively small DC motors rotated the braced thigh forward on one side while another motor rotated the side backward. The resulting reciprocal movement could also be controlled by a switch in the shoe, or as I called it, "brain in the shoe." The quadriceps was easily stimulated electrically and the foot could be either activated by the peroneal brace as I described above or be braced.

In a paper presented at the International Meeting in Puerto Rico in 1966 I offered a blue print for further developments of electrology applied to rehabilitation medicine. In this paper I stated that the hip mechanism is easily realized by electric motors driving hip or long leg braces (articulated with a pelvic band or a pelvic corset) in opposite directions on each side being automatically controlled from the switch in the shoe. It is interesting to note that this paper in 1966 was presented a century after the publication of a famous book by Raymond DuBois, who initiated electrical stimulation of the muscles, demonstrating their function, although he did not use it for actual activities of the individual in daily life.

A problem arose however inasmuch as if the patient did not have functional hysteria, the foot that could help in some individuals was helplessly stopped by friction of the floor.

Seven years after this paper, I found the solution to this problem in a paper presented to the Academy and Congress of Rehabilitative Medicine in 1973. I reported my findings of the possibility of "reflex walking" in paraplegic patients. I found that in such patients a low intensity of 40 cycles per second, 50 to 100 microsecond electrical pulses applied to the sural or tibial nerves (in the popliteal fossae) result in reflex hip

Functional Electrical Stimulation

and knee flexion with a simultaneous reflex ankle dorsiflexion, simulating a swing phase of the gait. This reflex proved to be somewhat capricious, yet in many patients it induced a significant shortening of the leg, thus permitting my mechanical systems of motors to be effective.

Thus, starting in 1960 and finishing in 1973, my system of standing and walking paraplegics was clearly formulated with no surgery being necessary for this system.

In this system the patient initiates a preprogrammed sequence of stimulae by a portable computer giving an order to stand, sit down, or walk. The patient is able to interrupt the sequence at any time.

It had been discovered that the stimulation of the agonist produces inhibition of antagonists and this system was used in many cases of spinal cord injury. I wondered whether my system could be useful for this application. I may add that on the basis of my experience, additional procedures can be used involving the latissimus dorsi. The latissimus dorsi is a long, powerful muscle which is innervated from the cervical spine and is attached to the illiac crest by the tendon. At least one of these muscles could be used for direct control over a muscle of the lower extremity.

In hemiplegic patients, the problem is quite simple as far as the lower extremities are concerned using the same principles. However, the upper extremity presents another problem. We showed that a few of these patients can benefit from electrophysiological techniques of muscle activation. Many of these patients were able to hold an object by their hypertonic finger flexors; however they could not release the grasp because of the hypertonicity of the finger extensors. We placed two electrodes under the finger extensors and by the operation of a micro-switch at the shoulder of the patient, the extensor could be contracted and the grasp released. We found however that this methodology was restricted to only a few patients because the electrical stimulation of the extensors produced a considerable decrease of spacicity and therefore the finger flexors were normal, adequate for the grasp. We stimulated the deltoid muscle producing a reduction of the subluxation where it existed, and the triceps and extensors of the forearm and wrist and fingers so that spasticity was reduced. Stimulation of finger flexors of the third dorsal interosseus may produce the grasp in cases where the patient can tolerate the stimulation. The procedure of surgery can be quite useful. We found that section of the most spastic muscle reduces the degree of spasicity in other muscles. On the other hand, by electromyography we find the muscle which has the best voluntary control, and this muscle can be switched by tendon surgery to become an active flexor of the fingers. If the muscle happens to be the extensor, we can activate it by electrical stimulation which is universally well tolerated by these patients. Adhesion is the greatest enemy of tendon sutures. I suggested "perpetual motion" of the fingers with a limited range realized by a variety of electrical and air-circulating devises acting around the clock.

Let us now look at other achievements of functional electrical stimulation. First, when spinal cord injury patients have difficulty in emptying their bladders, a system of stimulation of the sacral root is commercially available. I saw a demonstration at a conference in New York where the patient publicly emptied his bladder in a container just by pushing a switch under his body. At the conference in Vienna, we were given a report on fifteen years of experience of this surgery which has been almost universally successful. In addition to emptying the bladder, some of these patients recover their erections.

In some cases of restored facial muscles, it is very difficult to obtain a synchronization of the smile on both sides when it is provoked by an emotional stimulus. In cases where I use my electronic switch, a synchronization can be developed.

Functional electrical stimulation can also be used in cardiac surgery. Years ago, I submitted to the Veterans' Administration, a project to use skeletal muscles for replacing deficient cardiac muscles. My project was turned down, although I understand that Dr. Tentarovich had the same idea at about the same time. The realization of this idea depends on the possibility of transforming a skeletal muscle into a cardiac muscle. The skeletal muscle is more or less fatigueable; the cardiac muscle must remain continuously active indefinitely. This was recently achieved by a methodology promising considerable functional development. One has to simply stimulate the skeletal muscle at a low rate.

Stimulation of the phrenic nerve for artificial respiration is the most durable example of functional electrical stimulation. Here again surgery must be performed in order to stimulate the phrenic nerve terminal branches. This methodology is commercially available. Another important area of application of functional electrical stimulation is in the field of hearing. A devise receiving complex signals resulting from human speech may be transformed into electrical pulses capable of stimulating the cochlea. This is now being done with partial success. In addition attempts are being made to stimulate the occipital lobe cortex in order to make blind people see.

Working with Dr. Julia Terzis, I experimented with what I call combined functional electromyography and electrical stimulation. I applied this methodology to two different areas; in patients with evulsive brachial plexus lesions and in those with facial palsy. I demonstrated that in those cases in whom Dr. Terzis was unable to restore the mechanogenesis of the biceps for example, she could restore a significant electrogenesis; enough to activate an electronic switch which I had developed in collaboration with my technical assistant many years before. When the patient tries to contract his inefficient bicep, he closes the switch by the potentials deriving from the partially restored biceps and this activates a motorized sling placing the hand at different levels. In many cases I succeeded to activate functional electrical stimulation by contracting a muscle at a distance.

Yugoslavian workers who acknowledged my priority, developed a four channel system of stimulation which I had suggested sixteen years earlier. These workers, the most persevering foreign scientists working in this area, presented a simple solution for about 10 per cent of paraplegics. Obviously, patients with excessive ulcers, contractures, incontinence and little ambition could not be helped by this system. However, contrary to my original plan, these investigators preferred to place switches in canes or crutches rather than in the shoe. They showed that in order to achieve satisfactory results, many months of stimulating the quadriceps was necessary in order to increase their strength and the patient had to be taught to stand first and then to walk. I believe that in view of the problems of "reflex walking" the addition of my DC hip motors is more acceptable.

A system used in New Orleans also reproduced my design partly. In this system, the reciprocal action on the hip joint is not made by the DC motor, but by electrical stimulation itself. For example, the hamstring is stimulated on one side and the resulting flexion pulls a cable which places the other hip in hyperextention and visa versa; stimulation of the quadriceps flexes the hip on the opposite side by the action of a cable. Thus another realization of the reciprocal motion which I proposed years before. Here again, as with the Yugoslavian system, four channel stimulation activates the quadriceps and hamstrings on both sides, the rest of the leg being braced.

A German system of eight channels was able to realize a complete solution of the surface stimulation without individually stimulated muscles.

Despite the high promise of the above it is still unusual to see plegic patients walking around with stimulators instead of walkers and other mechanical prostheses, even in the nineties.

Electrodiagnosis and Electric Therapy

I was initiated into electrodiagnosis in the strictest meaning of the term, in the laboratories of Professor Lapicque and Doctor Bourguignon in Paris, the most outstanding protagonists of electrodiagnosis of the century. Lapicque stressed "time factors" in "Electrodiagnosis and Neurophysiology," his major book. He failed to convince us that it was the notion of chronaxie that was the main, useful expression of these factors. A few years later, Erlanger and Gasser in the United States demonstrated that time factors were most significant, differentiating various nerve fibers by measuring related conduction velocities, and won the Nobel Prize for their finding. We now know that conduction velocity is the most important single factor in differentiating pathology from normality, conduction velocity that was correctly measured in the laboratory of Helmholtz more than a century ago. Recent developments of EMG may be viewed in great part as an extension of conduction velocity techniques; first from motor to sensory fibers, then proximally and upwards, towards the plexuses, roots, spinal cord, brain stem and cerebral hemispheres. On this invasive path H-reflexes F waves, blink reflexes, silent periods, evoked potentials were used and now the R2 waves of Dr. Eisen. On the invasion path from the feet, pudendal nerves were given well deserved consideration. Dr. MacLean showed how one may stimulate roots and intra-abdominal nerves for this purpose.

There is however, a domain in which a more distinct application of the ideas of Lapicque and Bourguignon should be made, namely in therapy. Lapicque showed that effective electrical pulses involve minimal energy when their duration is equal to their chronaxie. This signifies a pulse duration of a fraction of a millisecond for innervated skeletal muscles. Bourguignon demonstrated that such pulses are the least painful. Neither Lapicque nor Bourguignon intended to use this notion in therapy. However, when I induced Doctor Offner to build me a brief stimulus machine for electric shock, at that time the first portable transistorized therapeutic stimulator for functional electrical stimulation and now used by thousands for TNS, I was guided by the ideas of Lapicque and Bourguignon which I verified experimentally. Such stimuli, as I showed, produced minimal skin irritation, minimal discomfort, maximum use of the battery, and therefore were of minimal cost to patients and society at large. Moreover, I demonstrated that such electrotherapy of innervated muscles increases their strength, decreases their fatigueability, spasms, spasticity, exerts effective traction on their retracted antagonists, increases blood circulation, and may induce useful function, so welcome to persons with central paralysis. Melzak and Wall implicitly followed directions previously suggested by Livingston and Beritoff. I translated Beritoff's book the same year of the publication of the paper by Melzak and Wall. I modified their techniques by placing electrodes at the axilla (brachial plexus) of popliteal fossa, in order to magnify the analgesic effects.

In the case of denervated muscles the lack of electrical stimulation may signify a failure to reinnervate distal muscles. Total denervation is relatively rare in practice, except in catastrophic nerve injuries. In cases of brachial plexus injury and nerve transplantation surgery, the absence of electrical stimulation is disastrous. Indeed, by the time there is successful reinnervation of a muscle, sometimes more than a year after surgery, some muscles of the forearm and all the intrinsics of the hand offer no muscle tissue to reinnervate. All experimental studies have shown that electrical stimulation conducted around the clock preserve muscle bulk for a year at least. Some animal studies have suggested a delay in reinnervation in electrically stimulated muscles. In a number of my patients who practiced electrical stimulation for 5 hours per day, reinnervation did take place. At any rate there is no choice in a case of a major gap between the place of injury and the muscles to reinnervate. Here again, stimulation has to be done with stimulation durations equal to the chronaxie of the stimulated muscle; however the chronaxie may be equal for denervated muscle, to 50, 70 or even 100 msec, to be determined individually and repeatedly during the course of therapy. With this duration of stimulation, skin redness will appear after only one half to one hour, even using one stimulus every 10 to 20 seconds, 3 times per minute, instead of 30 per second as in the case of innervated muscles, in order to reproduce a fused tetanus. In such cases one must secondarily interrupt the current. We use 23 seconds on and 2 seconds off for innervated muscles.

Coming back to the denervated muscle one must wait for the skin redness to disappear before starting a new session of therapy. We place the electrode above the most proximal muscle to stimulate, and below the most distal muscle. Nearly all muscles in between will be activated by this longitudinal stimulation. When the muscle becomes reinnervated brief stimuli may be used, preferably with voluntary contractions to help the current or with brief isometric exercises executed simultaneously. Some patients perform 1000 exercises a day at home. In children other exercises should be performed with the normal extremity bandaged to the body, using "bribes" to induce use of the involved extremity.

I shall discuss electrotherapy in anatomical order from the face to the toes except that in the case of the face, I use long duration stimuli, while my initial discussion is related to the use of brief stimuli. There are 3 main types of stimuli used in electrotherapy, the rest being the result of the imagination of enterprising manufacturers. One, a brief stimulus generally below a third of a msec. duration; two, a slower pulse with progressive onset, remembering that abnormally innervated muscles cannot be stimulated with progressively increasing stimuli rates; and three, slow pulses generally not exceeding 300 to 500 msec duration. We shall start with brief stimuli.

One of the most dramatic examples of therapeutic electrical stimulation occurs n acute torticollis, where a patient wakens with an ability to rotate his head on one direction because of pain and muscle spasms. You may also be dealing with patients in which the torticollis is chronic. I have demonstrated in both classes of patients that a strong single stimulus applied to the sternocleidomastoid muscle, that rotates the head in a non-painful direction, may be instantaneously curative (within seconds). The patient may have residual neck pain that one treats by stimulating the trapezius muscles with brief stimuli; the best you can offer the patient, except in cases of radiculopathy. A second still more common case is that of the patient who develops a restriction of shoulder abduction. If not treated it often leads to a "frozen shoulder." I treat such patients as follows: negative electrode to axilla, positive electrode to deltoid, hand on the wall. The third most common case is stimulation of the deltoid which is paretic in hemiplegia; this generates pain in the shoulder. Every physician should know that chronic stimulation of the deltoid restores the situation and suppresses pain in the shoulder.

Another common case is the so-called tennis elbow involving extensors of the wrist. Stimulation of these muscles is immediately helpful, although the condition tends to recur. A further common cause is the reduction of the range of motion at the wrist or fingers following fracture, or application of a cast. As in other cases the stimulation is daily and prolonged for hours per day leading to increase of range of motion unless there is a nerve injury. Finally, in hand pain or paresthesias resulting from many causes—radiculopathy, neuropathy, arthritis, carpal tunnel syndrome, etc., stimulation is applied to the median or ulnar nerves unless surgery is indicated. Before leaving the upper extremity, let me say that in hemiplegic patients, stimulation of the extensor mass of the forearm combined with stimulation of the triceps always decreases or suppresses spasticity.

So much for the upper extremities. Let us now consider low back pain. Of course the EMG and MRI have already established a radiculopathy, which may or may not require surgery. Otherwise, stimulation of the paraspinal muscles always decreases pain during stimulation. Moreover, because electrical stimulation exercises the muscle, the physician can apply a corset without fear of weakening these muscles. I am not saying that electrical stimulation is the only treatment for back pain. In certain cases one must advise surgery, in certain others one may use injections of ameliorating medications, but in all cases electrical stimulation of paraspinal muscles is helpful.

Let us now go to the lower extremities. Unfortunately it is rare that electrical stimulation is fully effective as far as the gluteus maximus is concerned. It is totally ineffective for the psoas, it is fully effective for the gluteus medius. It is most useful for the quadriceps. In order to stimulate the quadriceps one must use large electrodes, and when I say large, they should be almost half the length of the thigh, the lower one being negative. Stimulation must be quite long and the patient should tolerate it. It should be constant in hemiplegics having difficulty in walking. The technique is almost the same for stimulating the hamstrings when necessary. As is known, stimulation may help paraplegics to stand and walk, in hemiplegics it suppresses foot drop. In cases of a partial rupture of gastrocnemius, stimulation is helpful. In cases of foot pain I always stimulate via the nerve points for the tibial, peroneal and sural nerves. In practically all instances, stimulation suppresses pain during and sometimes after the stimulation. It is obvious, however, that other more specific therapies must be used in addition. As you can see from this cursory review the number of cases requiring stimulation is considerable and therefore electrical stimulation, prolonged and well applied can be used predominantly in a great number of cases.

So much for electrodiagnosis and electric therapy. We will now consider therapeutic electromyography or biofeedback. In patients whose nerves are diseased the motor units that they have been using all their lives may have been stricken by the injury to the extent that they are unable to initiate a movement. There are other patients whose nerve repair caused the proximal fibers to regenerate into channels of distal fibres that were destined to activate a completely different motion from that initiated by the brain through the proximal fibers. It is the task of the physiatrist, acting as a clinical neurophysiologist to guide this patient in finding effective motor units that will, under the new circumstances, bring about a desirable motor act. The physician does this by demonstrating visually on a cathode ray oscilloscope, and auditorily through a speaker system, which motor fibers are recruited by different motor commands, particularly those muscles that are located near the electrode tip. In this way the patient is helped in relearning motor commands. Those of us who have witnessed a patient with Bells Palsy coming to the examination room with a widely open eyelid, and leaving with a normally operating eyelid, will forever appreciate the use of biofeedback. However, it has to be done correctly, with needle electrodes for the most part, balanced amplifiers to minimize volume conduction and with visual display. Biofeedback can also be used effectively in motor unit disease.

It may be a good time now to lay the foundations of diagnostic electromyography. Years ago, as editor of the Bulletin of the American Association for Electrodiagnosis and Electromyography, at the occasion of 100 years of electrophysiology, I prepared an article. I was reminded that a century ago, Wedensky, in the laboratory in which I started my career as a neurophysiologist, listened and made others listen to a telephone which was connected by needles to muscles whose activity would be represented by noises perceived in the telephone. Whether this was an authentic electromyogram or an artifactual noise is really not important. The essential thing is that more than a century ago the idea of listening through microphones to electrical activity of muscles may have been hampered only by the deficiency of a technique developed not by neurophysiologists but by electrical engineers. But this has been the history of neurophysiology. As soon as a physical measuring or stimulating device was manufactured, it was applied to patients for potential therapy or for potential diagnostic functions. The essential notion in electromyography was that of the motor neuron and motor unit potentials. This was elaborated by Sherrington to whom we are obliged by so many other formulations in neurophysiology. It is important to stress that a single man in the span of one lifetime was able to illuminate so many aspects of the nervous system that his influence remains ineffaceable from the history of our science. Sherrington clearly formulated the notion that the motor cell in the anterior horn of the spinal cord gives rise to a considerable number of ramifications to its axon in such a way that each ramus will supply an individual muscle fiber. The number of muscle fibers supplied by one single axon, and therefore one single motor neuron may be counted in thousands for leg muscles, but only in units for supplying muscles of the arm. The gastrocnemius intervenes in voluntary activity in order to preserve posture, while the muscles of the eyes must participate and coordinate very delicate movements by which we are apprised by the distal receptors of minimal changes in the environment. Between these two extremes there are all sorts of muscles in which motor units contain all possible numbers of muscle fibers. This means that if voluntarily, one activates one cell in the anterior of the spinal cord, one brings about a contraction, sometimes of thousands of muscle fibers. Or if one succeeds in stimulating electrically one single axon,, the result will be the same. Each of the muscle fibers will produce an electrical potential. Now that we have learned to record with microelectrodes we know that this electrical potential is extremely brief in duration, from fractions of a millisecond to no more than one millisecond. The amplitude of the potentials is dependent on the distance of the recording electrode from the generating source.

Gone are the days when fibrillations were considered the main sign for differential diagnosis between neuropathies and myopathies on the one hand, and upper motor neuron disease on the other. They occur in all three conditions each time the axonic cytoplasm is affected. I showed that in hemiplegia they are much more prevalent in the distal than in the proximal muscles because the protoplasmic flow slows with distance. Also gone are the days when it was sufficient to make a diagnosis by listening to auditory signals. Computers help us now to practice quantitative EMG, in particular for the analysis of motor unit durations, another manifestation of the importance of Lapicque's time factor in electrodiagnosis. And gone are the days when we had jitters when we heard for the first time the term "jitter." Under the direction of Dr. Stalberg, an new single fiber EMG emerged out of this notion, useful not only for the diagnosis of myesthenia syndromes, but also for better follow-up of reinnervation. The greatest progress has been accomplished in studies of conduction velocity as already mentioned These studies have been crowned by the advent of averaging techniques.

However, let us first address a different aspect of the problem. Because of methodological difficulties, electromyographers have preferred to consider the fastest conduction velocities to characterize a neuropathy. It should be obvious that consideration of the whole scatter of conduction velocities within a nerve trunk would permit one to have a more exhaustive appreciation of pathological changes. Pending further progress in this direction, initiated by Professor Buchtal in the sensory nerves, one should at least consider the slowest fibers, non-myelinated fibers, so prevalent in our nerves. I met with them more than 30 years ago. As an electroencephalographer I was bothered by the artifacts caused by perspiration. I then developed a methodology to study conduction velocities in the corresponding fibers by simultaneously recording skin potentials from the scalp, hands and plantar regions of the feet just as is done on the hand by a "lie detector." It is helpful however, instead of inducing the patient to tell lies, to apply an electrical stimulus to any nerve or muscle. Thus a conduction velocity of 1 to 2 meters a second could be easily determined. A rather nice display of this phenomenon was with the neon toposcope of Robert Cohn. A clap of the hands showed flashing of a head lamp which was followed by successive flashings from the caudally placed electrodes. In the involved nerves, pathology manifests itself not by a slowing of conduction velocity but by a decrease of collapse of the amplitude of the skin potential.

Another and more colorful way of studying these nerve fibers is by thermography. Its application to the diagnosis of radiculopathies is still controversial. Its use in medico-legal cases may be quite important, particularly so in peripheral nerve injuries. I reported at the last International EMG convention a new method of "functional thermography." The temperature of the palm decreases in normals following peripheral nerve stimulation. Diagnostic applications of this observation remains to be studied.

In 1962, working with my Ph.D. student, Dr. Kim, I was impressed by a much lower amplitude component of the somato-sensory potential. Considering only negative potentials over 60 different points from which we made our records, we tracked these early low amplitude potentials and found them at their maximum amplitude over the neck region. We determined their latency to be 12+ or -1 msec. As noted earlier, I induced Dr. Voris, a neurosurgeon to track these potentials inside the spinal cord and midbrain where they had an additional 1 msec latency. In a later publication we suggested that thalamic potentials have a latency of around 15 msec. A few years later at Hines, I presented with my associates from the Department of Physical Medicine, a paper describing a technique of recording from the dorso-lumbar skin potentials arising from the roots and spinal cord. All these data were fully confirmed by Dr. John Desmedt of Belgium.

I am sure it is difficult for my contemporary young colleagues to imagine what kind of abuse we had to endure from those who would not believe in the veracity of our findings. Later, Dr. Cracco, working in Dr. Bickford's laboratory confirmed my findings. The results were com-

Functional Electrical Stimulation

pletely decisive. Dr. Cracco, in his enthusiasm for the early potentials, chose to downgrade the later components of the somato-sensory potentials. Although I am grateful to him and his wife for their work, which continued what I initiated, I disagree with him as to the clinical significance of the components. Only a few years after our discovery of the early potentials, I showed that not only the latency but the amplitude of the later potentials are clinically important. Thus we showed that severely involved hemiplegics, particularly those with aphasia, may show complete collapse of the amplitude of the late component sensory potentials. Recently Dr. LaJoie and then Dr. Alfred Pavot confirmed our findings and showed their importance as predictors in rehabilitation. More recently Dr. LaJoie and I showed that this amplitude collapse related to the component N20, may be present in the parietal cortex and not in the frontal. Prognostic significance of coma following severe brain lesions has been seen by many investigators.

The question remains: why, despite what I have just reviewed, is stimulation not the treatment of choice in most cases? The answer is quite simple; most patients dislike electrical stimulation, which is painful when applied with stimulators presently on the market. In order to appreciate the problem, one has to keep in mind the fundamental law of electrical stimulation expressed by strength duration curve discovered by Hooweg at the beginning of the century. This is the law. Within certain limits the decrease of pulse duration requires a corresponding increase in voltage. In order to utilize effective stimulation one makes the pulse duration as brief as possible. Why? Again it was discovered in the 30's that both sensory and motor fibers are of different diameters. The small diameter fibers when stimulated by slow pulses transmit pain and temperature predominantly, while fibers of large diameter stimulated by brief pulses transmit all other perceptions, as well as motor commands. It therefore makes sense to stimulate the nerve with the briefest possible pulses, so that the chances of transmitting pain are decreased while the transmission of motor commands is preserved. However, we know that when we use ultra brief stimuli we have to increase the voltage. Would this increase of voltage be possible without reintroducing pain transition? The only way to judge was to try, and so I tried.

I developed a micropulse stimulator using a voltage of about 300 volts which permitted the use of stimuli as brief as 10 microseconds. Our testing of these stimulators showed that the patients were able to tolerate these stimuli much better than any commercially available instruments, with the therapeutic results fully preserved. As a matter of fact even these brief stimuli appeared disagreeable to some patients but the number of super-sensitive patients is relatively small. As you can imagine this simple change permitted me to expand considerably the application of electrical stimulation. One must remember that such stimuli are effective only in those patients in whom the muscles remain innervated by the myelinated fibers. In case of a total nerve injury stimulation is still useful, but only in so far as preserving the muscle mass to be reinnervated in the future. Microsurgery also increases considerably the probability of such reinnervation.

The main requirement is to use stimuli long enough to activate denervated muscles on the order of 100 to 300 microseconds. These pulses may not be too painful unless the pulse duration is very prolonged. Yet the only portable commercial stimulator on the market in this country is the one in which each stimulus is dispatched by the patients themselves. Opening and closing a switch takes more than a second and therefor is too painful In order to reduce the pain the patient reduces the current making the procedure inefficient. I therefore developed an experimental stimulator that permits the use of pulses of only 100 to 300 milliseconds duration. Unfortunately these repeated pulses, without damaging the muscles, may burn the skin. After seven years of study we established that burning of the skin depended on the location of the electrode. With use of the properly positioned electrodes, the skin of the face, the shoulder, the palmer aspect of the hand and the plantar region of the foot as well as the skin of the dorsal thigh and that covering the extensors of the hand are less likely to burn than nay other regions. This seems to reduce the application of electrostimulation to only restricted regions of the body. Facing this difficulty we found an answer by remembering the principle of a potentiometer. One applies the positive potential to one end of the circuit and the negative to the other. Then by drawing the current away from any point of the resistor one may obtain currents of various intensities. I reasoned that one could consider the upper extremity as a potentiometer and bring one pole of the current to the shoulder and the other to the hand. Then any muscle in between will derive push of the current which is sufficient to activate this muscle automatically. Experimentation verified my prediction. For example, in a case of a total paralysis resulting from a brachial plexus total injury, the application of the current to the shoulder and the head of the deltoid will result in contraction of muscles in between, the deltoid, the biceps, extensors and flexors, as well as the intrinsics of the hand. While physical therapists stimulate separately these muscles by putting electrodes on them, they immeasurably decrease the time of stimulation of each of them. I stimulate all of them at once by my technique. The same is true, of course, of the leg. And so by this methodology and by using only necessary stimulus durations I have increased considerably the efficiency of electrotherapy in these cases.

One more problem. Suppose that an abnormally denervated muscle is located next to a normal muscle. Under these conditions one applies unnecessarily long duration stimuli to the normal muscle. How to prevent this? Simple. One must remember that normal muscles are not stimulated by progressive currents. Yet the latter are fully effective for denervated muscles. I therefore introduced progressive current in my stimulators. In a case of facial palsy when the current is applied to the denervated face muscle it will also stimulate the adjacent normal masseter muscle, which requires no stimulation. By using progressive currents, the masseter is not stimulated. It remains to determine the necessary time of application of therapy under different circumstance. Despite the available surgery concomitant with electrotherapy, very few surgeons who happen to injure the facial nerve in patients with acoustic neuromas advise the patient to have additional microsurgery and therapy for the injured facial nerve. And yet this should be done.

In the meantime we continue to stick to our immediate purpose; to contribute to the rehabilitation of paralyzed individuals. In the case of paralysis resulting from brachial plexus lesions, new intraoperative techniques are about to emerge; e.g., recording of muscle potentials simultaneously with those of somato-sensory potentials, while one stimulates one after the other from different roots. This method is bound to facilitate localization of lesions. In terms of rehabilitation of these patients, already simple devices may spell out useful functions for patients.

Newly introduced techniques by Dr. Cracco of brain stimulation by safe currents is another step forward. Dr. Terzis' technique of nerve transplants is bound to initiate a revolution in rehabilitation surgery in spinal cord injury and hemiplegic patients. On the other hand, some of the previous applications of functional electric therapy could not be realized because of the discomfort to the patient. The possibility of surgical desensitization of the area involved will undoubtedly permit one to significantly increase the scope of functional electrical stimulation.

Despite all the foregoing, I recently had the opportunity of talking to a number of residents in PM&R and found not only an almost complete ignorance of the principles of electrotherapy but also a feeling that there was nothing to learn about it. All that is needed for an occasional patient complaining of pain is to ask a therapist to arrange for the purchase of a "transcutaneous" stimulator. Everyone should be aware that all electrical stimulation conducted in a department of physical medicine is transcutaneous. I therefore welcome the opportunity to share the results of my practice of electrical stimulation, which I firmly believe is the single most important tool of physiatrists to help their patients. As you have seen, I use therapeutic electrical stimulation not only for pain but also to increase strength and endurance of muscles still being innervated by the spinal cord; or to contribute to their regeneration in case of total nerve injury. As you can see in most of my applications the "gate theory" has no place, as the improvement is usually secondary to the activation of the muscles themselves.

In conclusion may I say this: Whatever will be accomplished now will be little in comparison with the progress to be expected in the future. So my younger colleagues will have a great deal to expect during their scientific lifetimes.

> *"Dead was the poet, his hopes, his dreams*
> *His hidden thoughts, words that he said*
> *Who will recall them and redeem*
> *His quest for truth. The poet was dead."*